T0226096

Introduction to Nuclear Reactor Experiments

Genichiro Wakabayashi · Takahiro Yamada ·
Tomohiro Endo · Cheol Ho Pyeon

Introduction to Nuclear Reactor Experiments

Genichiro Wakabayashi
Atomic Energy Research Institute
Kindai University
Osaka, Japan

Takahiro Yamada
Atomic Energy Research Institute
Kindai University
Osaka, Japan

Tomohiro Endo
Graduate School of Engineering
Nagoya University
Nagoya, Japan

Cheol Ho Pyeon
Institute for Integrated Radiation
and Nuclear Science
Kyoto University
Osaka, Japan

ISBN 978-981-19-6591-3   ISBN 978-981-19-6589-0 (eBook)
https://doi.org/10.1007/978-981-19-6589-0

Translation from the Japanese language edition: "Genshiro Jikken Nyumon" by Genichiro Wakabayashi et al., © Genichiro Wakabayashi, Takahiro Yamada, Tomohiro Endo and Cheol Ho Pyeon 2022. Published by Kyoto University Press. All Rights Reserved.
© The Editor(s) (if applicable) and The Author(s) 2023. This book is an open access publication.
Open Access This book is licensed under the terms of the Creative Commons Attribution-NonCommercial-NoDerivatives 4.0 International License (http://creativecommons.org/licenses/by-nc-nd/4.0/), which permits any noncommercial use, sharing, distribution and reproduction in any medium or format, as long as you give appropriate credit to the original author(s) and the source, provide a link to the Creative Commons license and indicate if you modified the licensed material. You do not have permission under this license to share adapted material derived from this book or parts of it.
The images or other third party material in this book are included in the book's Creative Commons license, unless indicated otherwise in a credit line to the material. If material is not included in the book's Creative Commons license and your intended use is not permitted by statutory regulation or exceeds the permitted use, you will need to obtain permission directly from the copyright holder.
This work is subject to copyright. All commercial rights are reserved by the author(s), whether the whole or part of the material is concerned, specifically the rights of reprinting, reuse of illustrations, recitation, broadcasting, reproduction on microfilms or in any other physical way, and transmission or information storage and retrieval, electronic adaptation, computer software, or by similar or dissimilar methodology now known or hereafter developed. Regarding these commercial rights a non-exclusive license has been granted to the publisher.
The use of general descriptive names, registered names, trademarks, service marks, etc. in this publication does not imply, even in the absence of a specific statement, that such names are exempt from the relevant protective laws and regulations and therefore free for general use.
The publisher, the authors, and the editors are safe to assume that the advice and information in this book are believed to be true and accurate at the date of publication. Neither the publisher nor the authors or the editors give a warranty, expressed or implied, with respect to the material contained herein or for any errors or omissions that may have been made. The publisher remains neutral with regard to jurisdictional claims in published maps and institutional affiliations.

This Springer imprint is published by the registered company Springer Nature Singapore Pte Ltd.
The registered company address is: 152 Beach Road, #21-01/04 Gateway East, Singapore 189721, Singapore

Foreword

Reactor experiment programs at the University Training Reactor, Kindai University (UTR-KINKI) have been implemented by a successful publication of a textbook for educational experiments. A noteworthy breakthrough of UTR-KINKI is, here, considered to be achieved by the textbook publication.

The beginning of UTR-KINKI dates back to 1959. At that time, the USA was trying to spread the peaceful uses of nuclear energy around the world, and the teaching reactor was exhibited at the Tokyo International Trade Fair in Harumi, Tokyo, where the reactor was also operated as a demonstration. Mr. Koichi Seko, the first president of Kindai University, was one of those who had a look around the reactor. He had his own opinion that "science should be put into practice." He also thought that fostering of professional engineers who would lead the nuclear industry in the future was an important issue. From the background mentioned above, he made the decision to purchase the reactor. In 1960, the Atomic Energy Research Institute (AERI) was established, and in 1961, the Department of Nuclear Reactor Engineering was established in the Faculty of Science and Technology. In accordance with the installation of the reactor, UTR-KINKI reached a first criticality on November 11, 1961, making the first private nuclear reactor and the first university reactor in Japan.

The UTR-KINKI has been used for the practical training of undergraduate students of nuclear engineering and other related fields in Kindai University. We have engaged in the operation of UTR-KINKI for a long time. At the time of this writing, there are only two reactors in Japan that can be used for nuclear reactor training: UTR-KINKI and Kyoto University Critical Assembly (KUCA). It is no exaggeration to say that the UTR-KINKI is a valuable educational resource for nuclear education in Japan and an asset of our country. Since 2013, UTR-KINKI has been used for nuclear education under the International Nuclear Human Resource Development Initiative Project of the Ministry of Education, Culture, Sports, Science and Technology in Japan (MEXT). Under the current education project founded by MEXT in 2020, we realized the "activity" of practical training as an experimental textbook. I hope you could make great use of this textbook.

In 1974, the full power of UTR-KINKI was increased from 0.1 to 1 W, keeping now the operation of 1 W. The UTR-KINKI helps beginners and the general public to understand nuclear reactors. The simplicity of the reactor body makes it easy to understand the structure and principle. Since no cooling is required due to the extremely low power, reactor safety is considered high. Moreover, because the amount of fission products is small due to the low-power operation, easy access to the core is possible immediately after the reactor shutdown. The size of the reactor is suitable to conduct reactor educational and training programs. All major objects can be seen in one view, giving a sense of intimacy. The experience of reactor operation gives a strong impression of realism and tension together with educational effects. The reactor has its own role to play. The fact that the full power is limited to 1 W supports markedly an actual achievement of education effects with the use of the reactor.

The cost and effort required for the maintenance and management of UTR-KINKI is not a small matter. All the process related to UTR-KINKI has been demonstrated by the background that Kindai University has succeeded the wishes of its first president and that all the staff members of AERI have faithfully engaged in their work. I would like to share the joy of the completion of this textbook together with the staff of the current and past generations.

In educational and training programs at UTR-KINKI, physical phenomena occurred in the reactor are taken as measurement results, which are then analyzed by continuing academic discussions based on the theory. The results of computational simulations provide a quantitative complement between physical phenomenon and elementary processes. In general, it is expected that academic textbooks will also evolve with the combination of the improvement of measurement methods and the development of simulation technologies. Not only with the progress of measurement methods and computational techniques, but also with the change of educational needs and the revision of nuclear related laws and regulations, the textbook will be revised in future.

Kindai University has newly established the Department of Energy Materials in the Faculty of Science and Technology and has accepted new students from April 2022. I would like to think, the expectation that this book will be published around the end of 2022 is some kind of good omen.

Finally, Prof. Cheol Ho Pyeon (Institute for Integrated Radiation and Nuclear Science, Kyoto University) and Prof. Genichiro Wakabayashi (AERI, Kindai University) played a leading role in the publication of this book. I would like to express my gratitude and appreciation for their efforts.

July 2022

Prof. Hirokuni Yamanishi
Director, Atomic Energy Research
Institute
Kindai University
Osaka, Japan

Preface

The foundation of the University Training Reactor, Kindai University (UTR-KINKI) was based on high expectations of the founder (Mr. Koichi Seko) of Kindai University for nuclear energy, and on a noble philosophy reminiscent of the enterprising spirit in the field of nuclear energy. Also, he had kept his mind on the future energy situation in Japan and the subsequent nuclear education at universities. At the establishment, UTR-KINKI stood alone on the university's vast campus in Higashi-Osaka City where Kindai University is located. Now is Higashi-Osaka City one of the largest commuter towns in Osaka Prefecture. Not many students and graduates of Kindai University know that there is a small nuclear reactor for educational purposes on the campus, partly because a reactor building is surrounded by residential areas. Moreover, it is not well known, even among those who are engaged in nuclear energy, that the operation, inspection and maintenance management of UTR-KINKI, as well as the administrative work related to the regulation of the reactor, have been supported for many years by its own expenses without depending on financial support from the government. Thus, there is no need to argue that the background of establishment of UTR-KINKI is very different from that of the national university and the national research institute that have nuclear reactor facilities: UTR-KINKI is a very rare nuclear reactor facility in Japan.

Since the establishment in 1961, UTR-KINKI has been used for educational reactor experiments and basic research in various fields related to the utilization of radiation, including reactor physics, radiation detection, radiation health physics, activation analysis, radiation biology, medical applications and archaeology. As an affiliated institute of Kindai University, the Atomic Energy Research Institute (AERI) is well known nationwide for educational activities focusing on the nuclear education using a nuclear reactor, and the AERI professors and technical staff members are engaged in academic and research activities in a wide range of fields related to nuclear energy.

For over 60 years since the establishment, AERI has focused especially on nuclear education using UTR-KINKI. In addition to a long history of the UTR-KINKI operation, AERI has gained extensive experience in educational activities for undergraduate and graduate school students in domestic and overseas, elementary school, junior high school and high school teachers, high school and junior high school students, and the general public. It is a great pleasure for me, as a professor involved in nuclear education, to have the opportunity to compile the valuable educational asset of reactor experiment programs at UTR-KINKI.

This book is an academic textbook on nuclear reactor experiments jointly written by Prof. Genichiro Wakabayashi (AERI, Kindai University, Japan), Prof. Takahiro Yamada (AERI, Kindai University, Japan) and Prof. Tomohiro Endo (Nagoya University, Japan). As experts in radiation detection, nuclear regulations and reactor physics, respectively, the three professors have devoted much time and effort to writing this book. The textbook covers almost all the experiments that can be carried out at UTR-KINKI, and will provide an easy understanding of nuclear reactor experiments for students who want to study radiation detection, nuclear regulations and reactor physics at universities. The target readership of this book is, however, different from that of its sister book, "Nuclear Reactor Physics Experiments" (2020, Kyoto University Press, Kyoto, Japan). The textbook is intended for third-year undergraduates who will study nuclear energy, whereas "Nuclear Reactor Physics Experiment" is designed for graduate students who have studied reactor physics and radiation detection beforehand in their undergraduate years. The authors hope that this book could be the best textbook for students exploring nuclear energy-related fields to understand fundamental theories and principles of several experiments in reactor physics and radiation detection.

The authors are thankful to Springer Nature for having the patience to complete the manuscript and their responses throughout its preparation for about a year.

Finally, it should be noted that the textbook was published with the subsidy of the project for "Establishment of Nuclear Education Platform Focusing on University Research Reactors and Large-Sized Facilities" of "Global Nuclear Human Resource Development Initiative in 2020" founded by the Ministry of Education, Culture, Sports, Science and Technology (MEXT) in Japan.

July 2022   Prof. Cheol Ho Pyeon
Institute for Integrated Radiation
and Nuclear Science
Kyoto University
Osaka, Japan

Contents

Chapter 1
Nuclear Reactor for University Education

Abstract The University Teaching and Research Reactor of Kindai University (UTR-KINKI) is an education-oriented research reactor with a rated thermal power of 1 W. The reactor has been used for nuclear education and research in Japan for more than half a century since its first criticality in 1961. The characteristics of UTR-KINKI are described with the structures of fuel, moderator, reflector, core components and nuclear instrumentation in the reactor. Basic concepts of reactor physics parameters are complementally provided by explaining neutron flux, cross sections, reaction rates, neutron multiplication and reactivity, to understand the neutron characteristics of the reactor. Also, detailed information on the UTR-KINKI reactor operation is uniquely included in this chapter to fulfill the objective of contacting the reactor itself, together with nuclear regulations concerning safety instruction, radiation control, physical protection and nuclear fuel materials in nuclear facilities.

Keywords UTR-KINKI · Core components · Neutron characteristics · Reactor operation · Nuclear regulations

1.1 Kindai University Reactor

1.1.1 Overview

The University Teaching and Research Reactor of Kindai University (former Kinki University) (UTR-KINKI) is an education-oriented research reactor with a rated thermal power of 1 W in the Atomic Energy Research Institute (AERI). UTR-KINKI has been used for education and training of the students specializing mainly in nuclear science and engineering, and many undergraduate and graduate students have participated in the practice and workshop not only from Kindai University but also from other universities in Japan. In addition, the reactor has been contributing to secondary education by holding workshops for science teachers and high school students. Recently, UTR-KINKI has been used for training of foreign engineers and researchers from several countries that are newly introducing nuclear energy, and for employee training for domestic companies in the nuclear industry. The UTR-KINKI

© The Author(s) 2023
G. Wakabayashi et al., *Introduction to Nuclear Reactor Experiments*,
https://doi.org/10.1007/978-981-19-6589-0_1

is also actively used for various basic researches, including reactor physics experiments, detector developments and biological irradiations, as a joint-use facility for researchers from all over the country.

1.1.2 History

The UTR-KINKI is one of the University Teaching and Research reactors (UTRs[1]) designed and produced by the American Standard Corporation for university education and training, based on the Argonaut reactor developed by the Argonne National Laboratory in the USA. The UTR was originally exhibited by the US Atomic Energy Commission at the 3rd Tokyo International Trade Fair held in May 1959, operating for 18 days at the fair. Mr. Koichi Seko, the first president of Kindai University, decided to purchase the exhibited reactor when he had a look around the UTR at the fair (Ref. [1]). Then, the UTR was installed on the campus of Kindai University as UTR-KINKI, reached a first criticality at 20:53 on November 11, 1961, and began operation as the first private nuclear reactor and university nuclear reactor in Japan. The rated thermal power at the start of operation was 0.1 W, which was increased to 1 W in 1974 (Ref. [2]).

The UTR-KINKI has been operated for more than half a century without any trouble, but was severely affected by the accident at Fukushima Daiichi Nuclear Power Station of TEPCO in March 2011. In response to the accident, the newly established Nuclear Regulation Authority (NRA) has developed the new regulatory standards [3] that tighten firmly nuclear regulations not only for nuclear power plants but also for test and research reactors. UTR-KINKI was also required to undergo a safety review to verify that it complies with the new standards, and it would not be allowed to operate until it completes the review. Although the UTR-KINKI was shut down for more than three years from February 2014, the UTR-KINKI passed all the examinations and inspections by NRA in March 2017 and resumed the operation as a first test and research reactor in Japan under the new regulatory standards.

1.1.3 Characteristics

The UTR-KINKI is a light-water-moderated and graphite-reflected heterogeneous thermal neutron reactor fueled by enriched uranium. The rated thermal power is only 1 W, and the reactor is operated as a zero-power reactor,[2] so the reactor is

---

[1] The nameplate attached to the main body of UTR-KINKI reads "University Teaching and Research reactor," and some documents refer to it as an abbreviation of "University Training Reactor" or "Universal Training Reactor".

[2] The reactor that is operated at a reactor power lower than 1 W with no coolant is called, "a zero-power reactor".

kept at normal pressure and temperature during operation and does not need coolant. The reactor has a simple and straightforward configuration with minimal reactor components such as fuel, control rods, moderators, reflectors and shielding. The reactor can be easily operated to start up, change power and shut down in a short time.

Main reactivity coefficients of UTR-KINKI are negative, including fuel temperature coefficient, moderator temperature coefficient and void coefficient. Therefore, the reactor has an inherent self-regulating capability to suppress the reactor power increase when the reactor temperature increases.

The total amount of fuel consumed so far since the start of operation in 1961 is almost negligible, because the amount of fuel consumed at the full power of 1 W is about 1 μg per day. Therefore, the fuel loaded at the start of operation is still in use, and there is no need to load new fuel in the future. For the same reason, fission products accumulated in the fuel are very small, so the radioactivity contained in the fuel is also small. Furthermore, because radiation leakage from the reactor is small, one can work in the vicinity of the reactor even while it is operated.

As described above, UTR-KINKI is extremely safe, easy to maintain, and produces almost no radioactive waste, making it suitable for education and research at universities. The main characteristics of UTR-KINKI are shown in Table 1.1, including the limiting conditions of reactivity.

Table 1.1 Main characteristics of UTR-KINKI

Rated thermal power		1 W
Annual integrated thermal power		1200 Wh
Average neutron lifetime		1.605×10^{-4} s
Excess reactivity		$< 0.5\% \Delta k/k$
Shutdown margin (with the most reactive control rod fully withdrawn)		$> 0.5\% \Delta k/k$
Negative reactivity due to experimental material inserted into the core		$< 0.3\% \Delta k/k$
Reactivity worth of experimental materials inserted into or taken out from the core during operation		$< 0.05\% \Delta k/k$
Reactivity worth of control rod	Safety Rods #1 and #2	$> 0.54\% \Delta k/k$
	Shim Safety Rod	$> 0.54\% \Delta k/k$
	Regulating Rod	$> 0.1\% \Delta k/k$
Reactivity coefficient	1% change in fuel mass	$\pm 0.33\% \Delta k/k$
	Temperature coefficient	$-0.008\% \Delta k/k \,°C^{-1}$
	Void coefficient of moderator	$-0.18\% \Delta k/k \,(\% \text{void})^{-1}$
Startup neutron source		Pu-Be 1 Ci ($1.4 \times 10^6 \, n \, s^{-1}$)
Maximum thermal neutron flux		$1.2 \times 10^7 \, cm^{-2} \, s^{-1}$

1.1.4   Configuration

The UTR-KINKI reactor consists of two fuel tanks containing fuel and water moderator, graphite reflector, four control rods, and five neutron detectors for nuclear instrumentation, and they are housed in the center of a cylindrical biological shielding tank. The core is a coupled core with a large space for irradiation in the graphite reflector between two fuel tanks. The reasons of this structure are to ensure that irradiation experiments are not restricted by water moderators and that the neutron flux in the reflector between the two tanks is flat so that inserting a large experimental sample in the center does not significantly affect the reactor characteristics. Photograph views of UTR-KINKI and its core are shown in Figs. 1.1 and 1.2, respectively, and top and side views of the UTR-KINKI core are shown in Figs. 1.3 and 1.4, respectively.

Fig. 1.1 Photograph view of UTR-KINKI reactor (© AERI, Kindai University. All rights reserved)

Fig. 1.2 Photograph view of UTR-KINKI core (© AERI, Kindai University. All rights reserved)

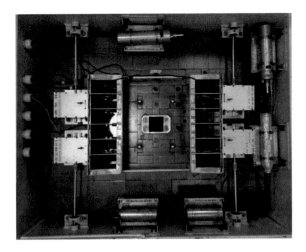

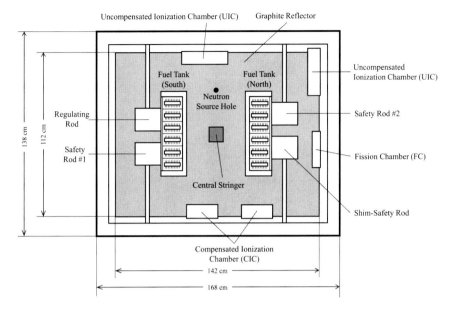

Fig. 1.3 Top view of UTR-KINKI core

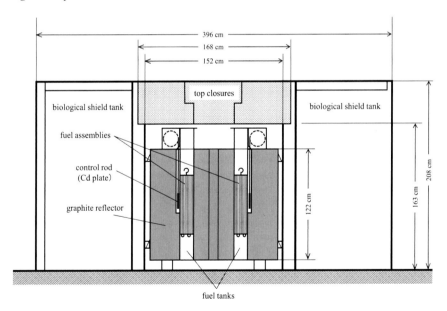

Fig. 1.4 Side view of UTR-KINKI reactor (North–South direction)

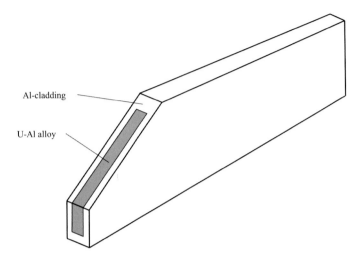

Al-cladding

U-Al alloy

Fig. 1.5 Structure of the fuel plate

1.1.4.1   Fuel and Water Moderator

The fuel of UTR-KINKI is a plate-type fuel made of uranium and aluminum (U-Al) alloy coated with Al clad. The fuel assembly consists of twelve fuel plates, and the single core contains six fuel assemblies. The moderator is light water, and the fuel assemblies in the two cores are always immersed in the light water. The structure of the fuel plate is shown in Fig. 1.5, and the detail of the fuel plate and the fuel assembly is presented in Fig. 1.6, and a model of fuel assembly is provided in Fig. 1.7.

1.1.4.2   Control Rod

The UTR-KINKI has four control rods, all of which use cadmium (Cd) plates as neutron absorbers (called "control rods," and, in fact, they are not rods but plates). The Cd plate moves up and down in the slit between the core and the graphite reflector. A conceptual diagram of the control rod drive mechanism is shown in Fig. 1.8. In most reactors, a rod-shaped control rod is inserted into the reactor, and the position of the rod is defined as the lower end of the rod. In terms of UTR-KINKI, however, the Cd plate moves up and down as a point-like neutron absorber. The range of the Cd plate movement is about 410 mm from the center of the fuel assemblies (lower limit) to the top of the graphite reflector (upper limit).

The Cd plate is attached to the end of a plate spring (stainless steel tape) and is withdrawn from the reactor by winding the spring with a rotating drum installed above the graphite reflector. The control rod drive motor is installed on the outer wall of the biological shielding tank and transmits power to the rotating drum through

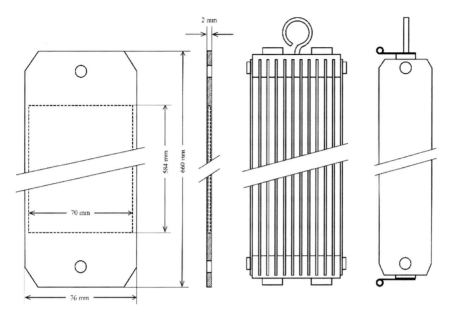

Fig. 1.6 Detail of fuel plate (left) and fuel assembly (right)

Fig. 1.7 Photograph view of fuel assembly (model) (© AERI, Kindai University. All rights reserved)

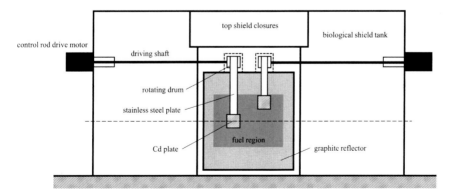

Fig. 1.8 Side view of operation of the Cd plate by control rod drive mechanism

a drive shaft penetrating the biological shielding tank and moves up and down the control rod.

Two of the four control rods are safety rods, which are called safety rod #1 (SR#1) and safety rod #2 (SR#2). The safety rods are used for emergency shutdown (scram) of the reactor and are withdrawn from the core at the start of operation. The Cd plates used for the SRs are 178 × 178 × 1 mm in size and have a large negative reactivity to shut down the reactor even if a single SR is inserted.

The remaining two rods are the shim safety rod (SSR) and the regulating rod (RR). The SSR and RR move up and down to control the reactivity of the reactor when the reactor is operated. The SSR is made of a Cd plate of the same size as SR and is a coarse regulating rod with a large negative reactivity. The RR is a fine-tuning control rod, and the size of the Cd plate is smaller (51 × 51 × 1 mm). Here, the reactivity of RR is about one-fifth of SSR.

When scram (emergency shutdown) is triggered in the reactor, the electromagnetic clutch of the control rod drive mechanism is disengaged. The SR#1, SR#2 and SSR make an immediate drop into the reactor by the restoring force of the plate spring and gravity. Since the three control rods are rapidly inserted, the reactor is then shut down within 0.5 s.

1.1.4.3 Graphite Reflector

The graphite reflector has a structure of graphite blocks assembled in a rectangular shape. When the graphite block called the "central stringer" in the central region of the core is pulled out, a vacancy for the central stringer can be used as an irradiation hole, which is often used in the irradiation experiments. Also, several other graphite blocks (called "vertical stringers") that can be pulled out are prepared in the graphite reflector. Some vertical stringers have multiple irradiation holes for setting samples to be activated and measuring the vertical neutron flux distribution. This allows foil activation analyses to be easily performed in training courses. A neutron source insertion hole and detector insertion holes are also provided in the graphite reflector.

1.1.4.4 Biological Shielding Tank and Top Shielding Closures

The biological shielding tank is a cylindrical tank, with a diameter of approximately 4 m and a height of approximately 2 m, filled with wet sand. During the operation, the top of the reactor is closed with top shielding closures made of concrete. In this way, radiations (neutrons and γ-rays) emitted from the core during operation are sufficiently shielded. Moreover, visitors can easily access to the area around the reactor even when the reactor is in operation.

Special experimental facilities can be installed on the top of the reactor in place of the top shielding closures. The A facility for small animal irradiation, B facility for neutron radiography or C facility for experimental sample insertion is selected and used according to the purpose of the experiment.

1.1.5 Nuclear Instrumentation

Nuclear instrumentation is an essential part of the equipment that measures the amount of neutrons in a reactor, a quantity proportional to the neutron flux, to obtain the information necessary for reactor control and protection and provide it to the operator. In general, the amount of neutrons varies over a very wide range (neutron flux ranging between 10^3 and 10^{14} cm^{-2} s^{-1} in a commercial power reactor) in reactor operation from startup to full power. Therefore, since a single neutron detector cannot cover the entire range, the measurement range is usually divided into three ranges (startup range, intermediate range, and power range), and information obtained from the detector in charge of each range is combined to obtain continuous information over the entire power range.

Since nuclear instrumentation is explained in detail in Chap. 5, Sect. 1.1.5 describes the minimal necessary contents of UTR-KINKI's nuclear instrumentation for preparatory study prior to participation in practical training.

The nuclear instrumentation of UTR-KINKI is shown in Fig. 1.9. UTR-KINKI has three nuclear instrumentation channels: the start-up channel; the intermediate channel; the power channel, and three neutron detectors are installed on the top of the graphite reflector for these channels. In addition, two neutron detectors for the safety channel are installed to trigger a scram signal.

A fission counter (FC) is used as a detector for the startup channel. The FC is an ionization chamber whose inner wall is coated with enriched uranium. When a neutron enters the chamber and triggers a fission reaction with the uranium, the produced fission fragments ionize the gas. The FC is characterized by its high neutron

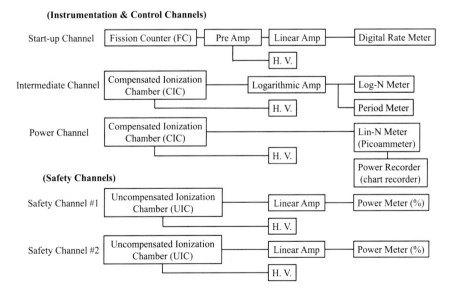

Fig. 1.9 Nuclear instrumentation of UTR-KINKI

sensitivity and a short dead time (about 10^{-6} s), and the neutron count rate is obtained in the pulse mode.

Compensated ionization chambers (CICs) are used in the current mode for intermediate and power channels. A CIC consists of two ionization chambers, one is sensitive to both neutrons and γ-rays and the other is sensitive only to γ-rays, and by taking the difference of the currents from the two chambers, a current proportional only to the neutron contribution is obtained. The inner wall of the neutron-sensitive ionization chamber is coated with boron (B), and charged particles (Li-7 and He-4) generated by the following reaction of B-10 and neutron ionize the gas:

$$^{10}B + {}^1n \rightarrow {}^7Li + {}^4He.$$

The ionization chamber, which is sensitive only to γ-rays, is not coated with boron.

The signal obtained from the CIC of the intermediate channel is processed by a logarithmic amplifier and then displayed on the logarithmic power meter (Log-N meter). The reactor period obtained by differentiating the signal from the logarithmic amplifier is also displayed on the period meter. The current signal from CIC in the power channel is directly measured with a picoammeter and displayed on the linear power meter (Lin-N meter). Furthermore, the signal from the linear power meter is continuously recorded on a chart paper by a power recorder (chart recorder).

Uncompensated ionization chambers (UICs) are used as neutron detectors in the safety channel. The UIC is an ionization chamber with a boron-coated inner wall. The UIC is also sensitive to γ-rays, so the output current includes the contribution of γ-rays. However, detectors used for the safety channel require more reliability than accuracy, so UICs are chosen for their simple structure. Two independent UICs are installed for multiplicity, and when the reactor power exceeds 150% of the rated thermal power (1 W), a scram signal is triggered and the reactor is immediately shutdown.

Other detectors for safety protection, not included in the nuclear instrumentation, are a seismic detector and a water level gauge installed on the side of the biological shielding tank. The seismic detector triggers a scram signal when it detects an acceleration of 100 gals or more, and the water level gauge triggers a scram signal when the water level in the biological shielding tank drops below 160 cm.

1.2 Characteristics of Neutrons in Nuclear Reactors

This section explains neutron characteristics that should be known when participating in nuclear reactor experiments, including neutron flux, nuclear reactions, cross section, reaction rates, neutron multiplication and reactivity.

1.2.1 Neutron Flux

Before explaining what neutron flux is, it is probably unfamiliar to those who are new to reactor physics. The closest image is magnetic flux in electromagnetics or heat flux in fluid mechanics.

To quantitatively express the distribution of neutrons in a nuclear reactor, we will use a quantity called neutron density. The neutron density, n, is defined as the "number of neutrons per unit volume: cm^{-3}." Considering the neutron that is on track to a specific direction at a certain speed ($cm\ s^{-1}$) in a particular material, if we focus on the number of neutrons passing per unit area (cm^2) and unit time (s^{-1}), the number of neutrons can be then expressed as follows:

$$n\left(cm^{-3}\right) \cdot v\left(cm\ s^{-1}\right) = n\ v\left(cm^{-2}\ s^{-1}\right). \tag{1.1}$$

The physical quantity expressed by Eq. (1.1) is called the "neutron flux $\phi\ (=\ n\ v)$."

On the other hand, how about observing the movement of neutrons with a different perspective as follows? Considering "the total moving distance of the neutron per unit volume (cm^{-3}) and unit time (s^{-1})," how can we express the neutron? The answer is the same as Eq. (1.1) (see [Column] "Some approaches to understanding neutron flux" in Sect. 1.2.3). From the viewpoint of reactor physics, the number of neutrons in a target object is the same whether it is a "surface" or a "volume." Let's remember that the concept of the "volume" is a different approach to understanding "neutron flux."

Additionally, the term "flux" has two definitions as follows:

(1) Vector quantity passing per unit area (cm^{-2}) and unit time (s^{-1}),
(2) Scalar quantity obtained by integrating the vector of definition (1) above in the angular direction per unit volume (cm^{-3}) and unit time (s^{-1}).

Finally, the definition of flux in engineering fields (electromagnetics, fluid mechanics, etc.) corresponds to the definition (1), and in reactor physics, the definition (2) is rather applied. Here, the vision that a "volume" is considered as a target of the neutron is convenient for understanding behavior of neutrons.

1.2.2 Cross Section

The term "cross section" gives an image of a side's area or a solid's cut surface, it is, however, difficult to be acceptable at face value. Like the neutron flux, the cross section is a physical quantity uniquely used in reactor physics (strictly speaking, nuclear physics). Since the cross section has the unit of the physical amount of cm^2, it is not entirely different from the image of the term. If you understand the physical meaning and phenomena, you will appreciate the consistency with the unit.

The cross section is a quantity corresponding to the probability of reactions between a neutron and a nucleus. Also, the cross section is a parameter for quantitatively treating physical phenomena occurring in a nuclear reactor. The cross section is determined by the relationship between the nucleus and the induced energy of the neutron. There are two types of cross sections: the microscopic cross section that is a physical quantity corresponding to the reaction of the neutron and the nucleus; the macroscopic cross section that is a physical quantity corresponding to that of the neutron and the material that constitutes one of reactor components.

1.2.2.1   Microscopic Cross Section

The interaction of a neutron and a nucleus (target nucleus) is stochastically triggered in a random process, and the "microscopic cross section" represents the ease of interaction of a neutron and a nucleus. Let's consider a nucleus as a target. Since the size itself of the nucleus is considered as microscopic cross section, the "microscopic cross section" has the unit of area (cm^2).

On the other hand, let us approach the "microscopic cross section" with a more concrete example. Imagine the following phenomenon using the velocity v (cm s^{-1}) and the neutron density n (cm^{-3}) defined in the explanation of neutron flux.

When a neutron is incident perpendicularly on a very thin material from one direction, let N (cm^{-3}) the target nucleus of the material. The number of collisions (the number of nuclear reactions) between neutrons and target nuclei is considered a criterion for the probability of the reactions and is taken as a proportionality constant σ. Since the number of collisions for the target nucleus is $n \cdot v \cdot \sigma$, number of collisions with the target nucleus per unit volume (cm^{-3}); that is, number of collisions R can be expressed as follows:

$$R\left(s^{-1}\right) = \sigma \cdot n\left(cm^{-3}\right) v\left(cm\ s^{-1}\right) N. \tag{1.2}$$

Considering R in Eq. (1.2) as the number of nuclear reactions per unit time, R is proportional to the number of incident neutrons, $n \cdot v$, and the number of nuclei in the target, N. As shown in Eq. (1.2), the proportionality constant σ has the unit of cm^2 that corresponds to the "microscopic cross section" and is consistent with the physical quantity corresponding to the reaction of the neutron and the nucleus. The unit of the "microscopic cross section" is 10^{-24} cm^2, called 1 barn.

1.2.2.2   Macroscopic Cross Section

On the basis of an easy understanding of the microscopic cross section mentioned in Sect. 1.2.2.1, you can interpret the "macroscopic cross section" as follows: the ease of interaction between neutrons and materials (not nuclei) is the macroscopic cross section. Also, the macroscopic cross section is expressed by multiplying the microscopic cross section (cm^2) by the atomic number density (cm^{-3}), with the unit of the inverse value of the length (cm^{-1}). The macroscopic cross section is understood as probability of interaction when a neutron moves through a material per unit length. Here, if the vision of understanding of "neutron flux" is applied to interpretation of macroscopic cross section, we can understand the macroscopic cross section physically as "the sum of the area of the target per unit volume $(cm^2 \, cm^{-3} = cm^{-1})$," because the macroscopic cross section has the unit of inverse value of length (cm^{-1}).

In this case, the macroscopic cross section Σ is defined by multiplying the microscopic cross section σ by the atomic number density N. In general, the macroscopic cross section is expressed as $\Sigma = \sigma \cdot N$.

Additionally, assuming that the total macroscopic cross section Σ_t is the sum of the macroscopic scattering cross section Σ_s and the macroscopic absorption cross section Σ_a (the sum of the macroscopic fission cross section Σ_f and the macroscopic capture cross section Σ_c), the inverse of the macroscopic total cross section L has the unit length (cm) and is expressed as $L = 1/\Sigma_t$. Here, L is called the mean free path of the neutron. The mean free path is interpreted as "the average distance taken from the incident that the neutron has a next reaction after causing a certain reaction with the material."

[Column] Relationship between cross section and energy

The cross section varies greatly depending on the energy of neutrons that cause nuclear reactions. This phenomenon is called the energy dependence of cross sections. Here, we introduce four (typical) characteristics of the energy dependence of cross sections (microscopic cross sections) as follows:

(1) $1/v$ characteristic (low-speed energy region)

The characteristic of the energy change in which the absorption cross section is proportional to the inverse of the velocity $1/v$ is called the $1/v$ characteristic. The characteristic is shown in a relatively low-energy region of the microscopic cross section of capture reactions. In particular, most of the nuclei with large mass numbers have the characteristic.

(2)   Resonance (medium-fast energy region)

The characteristic of a rapid change of cross section in a narrow energy range is called resonance. Such an energy change is caused by the existence of "excited energy level" inside the nucleus due to the shell structure of the nucleus. For neutrons at the energy that is consistent with the excitation energy level, the nucleus is very reactive with neutrons. A typical example of the characteristic is the resonance of U-238 at 6.7 eV.

(3)   Threshold reaction (fast energy region)

Neutrons sometimes cause nuclear reactions at the energy higher than a certain level. The cross section is then zero at the energy lower than a certain level and increases rapidly at the energy higher than a certain level. Such a reaction is called a threshold reaction, and the boundary energy at which the reaction causes, i.e., the energy at which the cross section is greatly large, is called the threshold energy. For example, the threshold energy of the neutron that causes fission reactions of U-238 is about 1 MeV.

(4) Others (wide energy range)

Depending on the neutron energy, the cross section may not change, which is often observed in scattering reactions. In this case, the cross section of the nuclear reaction is termed "a flat cross section."

1.2.3 Reaction Rate

We believe that you have a good understanding of neutron flux and macroscopic cross section, and we will define the physical quantity "reaction rate," which we want to focus on in this section. In Sect. 1.2.2, the cross section is interpreted as the ease (probability) of interaction of a neutron and a nucleus (target) or a material. Also, the neutron flux is conveniently taken as a group of neutrons. The rate that a group of neutrons interacts with a certain material is then called the "reaction rate" and expressed by multiplying the neutron flux ϕ (cm^{-2} s^{-1}) by the macroscopic cross section Σ (cm^{-1}). On the basis of the same concept shown in Eq. (1.2), the reaction rate R can be rewritten as follows:

$$R(\text{cm}^{-3}\ \text{s}^{-1}) = \Sigma(\text{cm}^{-1}) \cdot \phi(\text{cm}^{-2}\ \text{s}^{-1}). \tag{1.3}$$

The reaction rate is interpreted as corresponding to "the probability (number of reactions) that a neutron (population) interacts with a certain material per unit volume and unit time."

[Column] Some approaches to understanding neutron flux

In this column, another attempt is made to understand the neutron flux with the use of "neutron density" and "neutron current." The two terms are found in the field of reactor physics.

(1) Neutron density and neutron flux

Neutron density is defined as the number of neutrons per unit volume. Here, the number of neutrons per unit volume at time t and spatial position $\mathbf{r}$ (x, y, z) is defined as the neutron density $n(\mathbf{r}, t)$ $(n\ \text{cm}^{-3})$. Next, the reaction rate is defined as the number that nuclear reactions are trigged per unit volume, unit time at time t and spatial position $\mathbf{r}$ (x, y, z), denoted $R(\mathbf{r}, t)$ (cm^{-3}). Let's think about the relationship between the neutron density and the reaction rate.

We consider a group of neutrons moving through a material in a certain direction with a constant speed v (cm^{-1} s^{-1}) and focus on the number of neutrons passing per unit area and unit time. Since the distance that a neutron can move at the speed per unit time is v (cm), the number of the neutrons that can pass through the surface at the speed corresponds to that of neutrons existing in a distance (cm) away from the surface. Therefore, the number of the neutrons at the speed that can pass per unit area and unit time is that of the neutrons that exist in a solid (v (cm^3)) with a height (cm) and a base of unit area (cm^2). Since the density of neutrons is $n(\mathbf{r}, t)$ (cm^{-3}), the number of neutrons passing through a certain surface per unit time and unit area is expressed by the product of the neutron density and the distance flown per unit time by the neutron, as follows:

$$v(\text{cm}^3) \cdot n(\mathbf{r}, t)(\text{cm}^{-3}) = vn(\mathbf{r}, t). \tag{1.4}$$

The reaction rate $R(\mathbf{r}, t)$ (cm^{-3}) can be written by multiplying the number of neutrons passing through a specific surface per unit time by the macroscopic cross section ($\Sigma = \sigma N$) can be expressed as follows, using the number of neutrons $v n$ ($\mathbf{r}, t$) passing through a specific surface:

$$R(\mathbf{r}, t) = \Sigma(\mathbf{r}, t) \cdot v n(\mathbf{r}, t), \tag{1.5}$$

where $v n$ ($\mathbf{r}, t$) in Eq. (1.5) corresponds to the amount of the neutron flux.

(2) Neutron current and neutron flux

Neutron current is a vector quantity of net neutrons passing per unit area and unit time for a specific direction, indicating the unit of cm^{-2} s^{-1}.

Here, neutron flux can be interpreted using the following two definitions:

- Total number of neutrons passing per unit area and unit time $(\text{cm}^{-2}\ \text{s}^{-1})$
- Total distance taken by neutrons per unit volume and unit time $(\text{cm}^{-3}\ \text{s}^{-1}\ \text{cm} = \text{cm}^{-2}\ \text{s}^{-1})$.

In both cases, the unit is the same as cm^{-2} s^{-1}, and the difference is whether the movement of the neutron is defined as "area traversed" or "distance taken in volume." Observing the actual movement of neutrons, the definition using "volume" is considered more appropriate, strictly speaking.

1.2.4 Reactions of Neutrons with Nuclei

The interactions of neutrons and nuclei can be divided into two main parts: absorption reactions, in which the neutrons are absorbed by the nuclei; scattering reactions, in which the neutrons are scattering by resulting from collisions with the nuclei.

Here, the question arises as to whether fission reactions induced by the neutrons are significant. Fission reactions are, however, considered one of the absorption reactions, as shown in Fig. 1.10. This section describes briefly main characteristics of fission reactions, capture reactions, elastic scattering reactions and inelastic scattering reactions among the interactions of neutrons and nuclei.

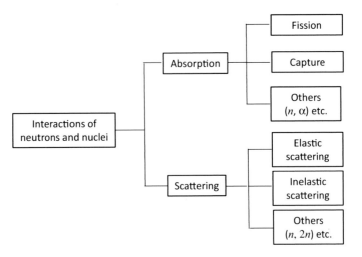

Fig. 1.10 Reactions between neutrons and nuclei

1.2.4.1 Fission Reaction

When neutrons collide with the nuclei of fissile materials (uranium, plutonium, etc.), fission reactions are induced with a certain probability, and the nucleus splits into two other nuclides, simultaneously releasing a large amount of energy and two or three neutrons. An example of fission reactions is shown in Fig. 1.11.

When a neutron is absorbed by the U-235 nucleus, a compound nucleus U-236 is generated after a short time, most of the U-236 nucleus splits into fission fragments, and two or three extra neutrons (average 2.5) are released. The neutrons emitted by

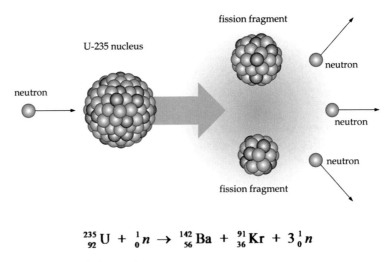

$$^{235}_{92}U + {}^{1}_{0}n \rightarrow {}^{142}_{56}Ba + {}^{91}_{36}Kr + 3{}^{1}_{0}n$$

Fig. 1.11 Example of fission reactions

fission reactions are called "prompt neutrons" with an average energy of 2 MeV. Since the energy of a particle is proportional to the square of its velocity, neutrons with energy comparable to MeV are called "fast neutrons" because they have high speeds.

Fission fragments produced by fission reactions move through the fuel for a distance of about 10 μm and then stop inside the fuel. Fission fragments stopped inside the fuel are called the fission products (FPs) that are highly unstable, because they are neutron-rich nuclei. Most of fission products emit one or more β^--ray and are stable nuclides. A small fraction of fission fragments emits neutrons, instead of β^--ray, making stable nuclides. The neutrons emitted are called "delayed neutrons," and the FPs that emit delayed neutrons are called "delayed neutron precursors." Again, the fission fragments produced by fission reactions are generally unstable because they have extra energy. Therefore, many fission fragments emit the extra energy as radiation, reducing their energy and making them at a more energetically stable state. That is why high radiation is emitted from nuclear fuel fissioned.

As mentioned above, when neutrons are absorbed by U-235 nuclei and fission reactions are induced, an average of 2.5 neutrons are emitted per fission. When one of 2.5 neutrons is absorbed by a next U-235 nucleus, further fission reactions are induced, and fission reactions continue in a chain reaction. A series of phenomena is called the "fission chain reactions." Some of neutrons generated by fission reactions are used to induce subsequent fission reactions. Meanwhile, other neutrons are absorbed by core components (e.g., control rods, fuel tank, and shielding materials) or leaked out of the reactor core without causing fission reactions in the reactor. The scheme of fission chain reactions is shown in Fig. 1.12. (For the sake of simplicity, the effects of neutron moderation and delayed neutrons are here omitted.)

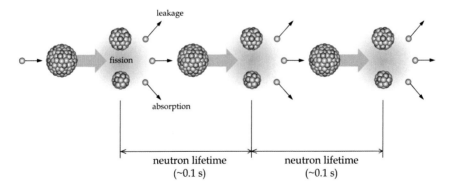

Fig. 1.12 Scheme of fission chain reactions

1.2.4.2 Capture Reaction

When neutrons are captured by collisions with nuclei, nuclei reach a state of increased energy caused by increasing the kinetic energy of neutrons and the binding energy of the nuclei. The phenomenon is called an excited state, and the nuclei in the state have extra energy. When the energy is released as γ-ray, the phenomenon is called "capture reactions."

In UTR-KINKI, control rods are made of Cd plates. Under the existence of Cd in the core, capture reactions are easily induced by the Cd absorption reactions with neutrons. Therefore, when control rods are inserted into the core, the neutrons in the core are absorbed by Cd, leading to a decrease in the total number of neutrons produced by fission reactions. Then, control rods play an essential role in reactor operation and power control, because control rods are responsible for determining the total number of neutrons produced by fission reactions and, consequently, for determining the reactor power.

1.2.4.3 Elastic Scattering Reaction

The elastic scattering reaction is a phenomenon in which a neutron is scattered by a collision with a nuclei. As the name implies, to understand what kind of phenomenon it is, it could be helpful to recall the diagram of the "law of conservation of momentum" that we took at lessons of physics in our high school days.

Let's consider the following two examples: the first example is a collision between a neutron and a hydrogen (H-1) nucleus. When the neutron and the H-1 collide head-on, the energy of the neutron is almost zero because the neutron and the H-1 have the same in weight, and the H-1 receives the energy of the neutron and carries out its motion. The second example is the collision between a neutron and a U-235 nucleus, where the U-235 weighs about 235 times as much as the neutron. Here, the U-235 keeps a steady state and the neutron is bounced back with the same energy. Notably, the neutron loses almost no energy at this point.

The elastic scattering reaction can be summarized as follows: when the elastic scattering reaction is induced by interactions of a neutron and a nuclei with a small mass number, a neutron loses a large amount of energy on average and is physically slowed down. On the other hand, when the elastic scattering reaction is induced by interactions of a neutron and a heavy nuclei with a large mass number, a neutron loses almost no energy; i.e., it is hardly slowed down. This is extremely important when understanding the multiplication of neutrons generated by fission reactions, so it is important to keep it in mind.

1.2.4.4 Inelastic Scattering Reaction

The inelastic scattering reaction is a phenomenon in which a neutron is absorbed by a nucleus, incorporating into the nucleus to form a compound nucleus. A particle (neutron) of the same type as the incident particle is then emitted from the compound nucleus. The kinetic energy of the incident particle is not conserved before and after the collision, because the kinetic energy of the incident particle is retained as the internal energy of the nucleus and is later emitted as γ-ray. The $(n, 2n)$ and $(n, 3n)$ reactions can also be considered as inelastic scattering reactions in a broad sense.

The inelastic scattering reaction should be higher than the minimum excitation energy of the nucleus; i.e., it has a threshold. For example, it is about 0.05 MeV for a heavy nucleus such as U-238 and more than 1 MeV for a light nucleus. Thus, inelastic scattering reactions are relatively high-energy reactions, which affect the formation of the energy distribution of fast neutrons in nuclear reactors.

1.2.5 Behavior of Neutrons

Based on the knowledge gained in Sect. 1.2.4, the behavior of neutrons in the core briefly describes in this section. The sequence of events is essential for understanding neutron multiplication later on.

Fast neutrons (prompt neutrons) generated by fission reactions repeatedly collide with light nucleus (hydrogen, oxygen, carbon, etc.) that constitute the moderators (light water, heavy water, graphite, etc.) in the reactor, and decrease their speed to make slow neutrons. After the neutrons are slowed down and steady in the low region, they are absorbed by U-235, and next fission reactions are then triggered. Here, fission chain reactions mean that a series of incidents: the generation of neutrons; slowing down of neutrons; absorption by U-235, is repeated continuously.

Slow neutrons are called "thermal neutrons," and their kinetic energy follows the Maxwell distribution. The peak energy is 0.025 eV, and the velocity is 2200 m s^{-1} (at room temperature). For example, in terms of a light-water-moderated reactor, the average time required from one fission to the next fission is about 0.2 ms when only prompt neutrons are considered. On the other hand, the average time is about 0.1 s when delayed neutrons are taken into account. The latter time is called the "effective lifetime."

[Column] Behavior of neutrons produced by fission reactions

This column introduces how neutrons generated by the U-235 fission reactions (see Fig. 1.11) behave in the core. The fact that neutrons generated by fission reactions contribute to the next fission reactions, which in turn lead to fission chain reactions, was explained in Sect. 1.2.4.1, "Fission Reaction." In this column, the interactions of the neutrons generated by fission reactions and the fuel are explained conceptually. You could then lead to an easy understanding of the concept of neutron multiplication in the core.

(1) Four-factor formula

Let's explain neutron multiplication using four factors (physical phenomena or interactions). First, we assume an infinitely large system consisting of only fuel such as

U-235 or U-238, instead of a complex system like a nuclear reactor. Then, the target neutrons are not neutrons provided from a neutron source, but neutrons generated by fission reactions. Here, we explain the phenomena (reactions) that can occur to a neutron.

Initially, a neutron is considered to be absorbed by the fuel in the system, and the probability is denoted by f (this is called the "thermal utilization"). The absorption is derived from the assumption that most fission reactions are induced by thermal neutrons. The probability that a neutron is absorbed by the fuel, and further fission reactions are induced is P_f. Here, when fission reactions are induced by neutrons, number of neutrons (ν) is generated. In this case, the average number of fission neutrons emitted per thermal neutron absorbed by the fuel is $\nu P_f (=\eta)$, called the "regeneration rate." Neutrons generated by fission reactions have an energy of 2 MeV, in which fission reactions by the interaction of U-238 and other materials are induced, and the ratio ε is called the "fast-fission factor." Finally, the fraction of 2 MeV fast neutrons that are not captured by resonance absorption and slowed down to thermal energy is p (called the "resonance escape probability").

The ratio of neutrons produced by the first fission reactions to those produced by the next fission reactions can be defined as the multiplication factor of neutrons. If we denote the multiplication factor as k_∞ for an infinite system, k_∞ can be expressed as follows, from the principle of superposition of the above four events:

$$k_\infty = f\,\eta\,\varepsilon\,p, \tag{1.6}$$

where k_∞ is called the infinite multiplication factor, and Eq. (1.6) is called the four factor formula for the infinite multiplication factor.

(2) Six-factor formula

In the four-factor formula, we consider that neutrons do not leak out of the system even if the system is finite, and the leakage is significant. If the probabilities of thermal and fast neutrons not leaking out of the system are P_T and P_F, respectively, Eq. (1.6) can be then expressed as follows:

$$k = f\,\eta\,\varepsilon\,p\,P_T\,P_F. \tag{1.7}$$

The multiplication factor k is called the effective multiplication factor, and Eq. (1.7) is called the six-factor formula.

The terms on the left and right sides of each group of the two-energy-group diffusion equation (see [Column] "Extension of the neutron diffusion equation to two-energy groups" in Chap. 2) can be regarded as the result of a simple expression of physical phenomena in the system of fast and thermal neutrons in Eq. (1.7). In other words, the six-factor formula explains qualitatively the behavior of neutrons in a finite system of nuclear reactors and appears to be a simple multiplication at first glance. Moreover, the formula is exquisitely conceived of the formula that accurately expresses the behavior of neutrons.

1.2.6 Prompt and Delayed Neutrons

1.2.6.1 Prompt Neutrons

Neutrons generated just after fission reactions are triggered are called "prompt neutrons," accounting for about 99% of the total neutrons generated in fission reactions. The number of prompt neutrons generated in fission reactions is statistically variable; some fission reactions may generate no neutrons at all, while others may generate as many as five neutrons. The average number of neutrons released per fission reaction is, however, needed for reactor calculations, which is about 2.5. New neutrons produced by fission reactions start next fission reactions at another location, and the behavior of the neutrons is continued one after another to start fission chain reactions. The average time between fission chain reactions (the time that neutrons generated by fission reactions in the reactor remain in the reactor until they are annihilated: the time interval between fission reactions) is called the "neutron lifetime" and varies depending on the reactor and fuel type, ranging between 10^{-6} and 10^{-4} s. The prompt neutron lifetime (ℓ) of UTR-KINKI is 1.605×10^{-4} s (160.5 μs obtained by MVP3.0 [4] together with JENDL-4.0 [5]).

The energy of the generated prompt neutrons is a continuous spectrum with an average energy of about 2 MeV, and the corresponding 2 MeV neutron velocity is 2 $\times 10^7$ m s^{-1}.

1.2.6.2 Delayed Neutrons

In addition to prompt neutrons (about 99%), delayed neutrons (about 1%) are generated in a short time, ranging between 10^{-2} and 10^2 s (about 0.7% in the case of thermal fission of U-235). Delayed neutrons are emitted as a product of β-decay when particular nuclides are produced among fission fragments (fragments produced by fission reactions), and the special fission fragments are called "delayed neutron precursors." As shown in Table 1.2, bromine (Br) (mass number: 87 to 92) and iodine (I) (mass number: 137 to 140) isotopes are the main nuclides of delayed neutron precursors. Delayed neutrons have an energy distribution with average energy slightly lower than that of prompt neutrons (about keV; prompt neutrons are in MeV) and have the same function as prompt neutrons in the reactor, except that the time to emission is longer (about 10^{-2} to 10^2 s) than that of prompt neutrons (about 10^{-7} to 10^{-5} s). Delayed neutrons are emitted later than fission reactions according to the half-life of the delayed neutron precursors. Moreover, delayed neutrons are generally divided into six groups according to their half-lives, as shown in Table 1.2.

Here, the role of delayed neutrons in a nuclear reactor is explained by the following simple calculation. Assuming that fission reactions are triggered and all the neutrons at that time are prompt neutrons, the time when prompt neutrons are generated simultaneously in fission reactions is about 10^{-7} to 10^{-5} s. Also, assuming that this time is about 10^{-5} s and that the number of neutrons in one fission reaction increases

Table 1.2 List of delayed neutron precursors

Group	Precursor	Half-life of precursors (s)	Average energy (keV)
1	Br-87	55.6	250
2	I-137 Br-88 Sb-134, Te-136, Cs-141	24.5 16.5	560
3	I-138 Br-89 As-84, Se-87, Rb-92, Rb-93, La-147	6.49 4.40	405
4	I-139 Br-9 Ga, As, Se, Br, Kr, Rb, Y, In, Sb, Te, I, Xe, Cs	2.29 1.92	450
5	Ga, As, Se, Br, Kr, Sr, Y, In, Sn, Sb, I, Xe, Cs, Ba	(~ 0.5)	–
6	Ga, Se, Br, Kr, Rb, In, Cs	(~ 0.2)	–

by a factor of 1.1, the number of neutrons after 10^{-3} s per 10^{-5} s increases by "a factor of 1.1 multiplied by 10^2 (= 14,000 times)." The multiplication rate is a tremendous increase in the number of neutrons.

Delayed neutrons, even if only about 1%, change the overall number of neutrons in the following way. The number of neutrons after 10^{-3} s for a time variation of every 10^{-2} s is "a factor of 1.1 multiplied by 10^{-1} (= 1.01 times)." The multiplication rate is a very small.

In the way, it can be understood that if delayed neutrons did not exist, it could be impossible to control the reactor stably with the time variation of prompt neutrons generated in fission reactions. On the contrary, the criticality of the reactor can be reached by relying on the time variation of delayed neutrons: by the time within a range of the half-lives of delayed neutron precursors.

1.2.7   Effective Multiplication Factor and Reactivity

Let's imagine the phenomena before and after a certain change occurs in the reactor, regardless of whether the reactor is at a critical, subcritical, or supercritical state, assuming that there is no neutron source in the reactor.

The ratio of the number of fission reactions occurring at a given moment (generation), f_1, to the number of fission reactions occurring after the neutron lifetime (the number of fissions after one generation), f_2 corresponds to k_{eff}. Then, k_{eff} can be expressed as follows:

Table 1.3 Relationship between effective multiplication factor, reactivity and reactor power

Effective multiplication factor k_{eff}	Reactivity	Reactor power change	State
$k_{eff} > 1$	> 0	Increase with time	Supercritical
$k_{eff} = 1$	$= 0$	Constant regardless of time	Critical
$k_{eff} < 1$	< 0	Decrease with time	Subcritical

$$k_{eff} = \frac{f_2}{f_1}. \tag{1.8}$$

In other words, k_{eff} is the ratio that indicates the multiple number of fission reactions increased in each generation. Since k_{eff} is close to 1 (unity) in a nuclear reactor, "the ratio of the deviation of k_{eff} from 1 (a quantitative index of the multiplication factor),"ρ, is used as an index of the deviation from criticality. Namely, when the condition of the reactor changes from a certain state (k_{eff}) to a critical state (effective multiplication factor is 1), the amount of change can be expressed by the difference between unity and the inverse value of the effective multiplication factor, as follows:

$$\rho = 1 - \frac{1}{k_{eff}} = \frac{k_{eff} - 1}{k_{eff}}, \tag{1.9}$$

where ρ is called the reactivity ($\Delta k / k$).

Since the power of the reactor is proportional to the number of fission reactions, a triadic relation between the effective multiplication factor, the reactivity and the change in the reactor power is shown in Table 1.3.

The control of a nuclear reactor is conducted by changing the value of the effective multiplication factor through the withdrawal or insertion of control rods. When the control rods are inserted into the reactor, the absorption rates of neutrons to the control rods increase, and the value of the effective multiplication factor is decreased. Conversely, when the control rods are withdrawn from the core, the value of the effective multiplication factor is increased.

As shown in Fig. 1.13, the reactor power with the effective multiplication factor of the reactor varies according to the time variation.

Here, we make an attempt to conduct a simple calculation. From Eq. (1.9), the reactor is assumed to have the positive reactivity of only $\rho = 0.03 \ \Delta k / k$ and the negative reactivity per each of three control rods is $\rho = -0.01 \ \Delta k / k$. When the three control rods are inserted fully into the reactor at a supercritical state with the positive reactivity of $0.03 \ \Delta k / k$, the excess reactivity of the reactor is then $\rho = 0 \ \Delta k / k$, and a critical state is maintained as shown in Table 1.3. Thus, the neutron balance can be obtained only by simple addition and subtraction using the positive and negative reactivities. The value of reactivity is an important index to understand the reactor state at this time.

Fig. 1.13 Effective
multiplication factor and
reactor power change

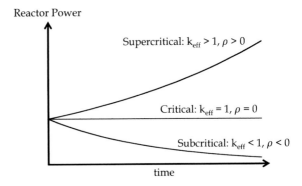

1.2.8 Reactor Kinetics

1.2.8.1 Prompt Neutron Lifetime

The time until the prompt neutrons generated by fission reactions are absorbed by the
fuel or annihilated by leakage is called the "prompt neutron lifetime" and is expressed
as ℓ_0. The prompt neutron lifetime varies depending on the type of reactor, and, for
UTR-KINKI, is obtained by using the Monte Carlo calculation code MVP3.0 and
the nuclear data library JENDL-4.0 as follows:

$$\ell_0 = 1.605 \times 10^{-4}\text{s}(= 160.5\,\mu\text{s})$$

1.2.8.2 Effective Delayed Neutron Fraction

In reactor calculations, it is convenient to use the ratio of the number of delayed
neutrons to the total number of neutrons (v_d) produced per fission reaction rather
than the absolute amount of delayed neutrons produced (v). The ratio of v_d and v
is called the "delayed neutron fraction" and is expressed as $\beta = v_d/v$. Similarly,
the ratio of the number of delayed neutrons in the i-th group precursor to the total
number of neutrons produced per fission reaction ($v_{d,i}$) is called the delayed neutron
fraction in i-th group and is denoted as $\beta_i = v_{d,i}/v$.

   The energy of delayed neutrons generated by fission reactions is smaller than that
of prompt neutrons, as shown in Table 1.2. Therefore, the delayed neutrons are less
wasteful than the prompt neutrons because the delayed neutrons leak out of the core
or are absorbed by core component materials in the process of neutron moderation.
In addition, the effective delayed neutron fraction β_{eff} that is specific to each reactor
is used in numerical calculations, because the neutron energy that fission reactions
are effectively triggered depends differently on the reactor type.

   The decay constants λ_i and effective delayed neutron fractions $\beta_{eff,i}$ of the
delayed neutron precursors in UTR-KINKI are shown in Table 1.4.

Table 1.4 Effective delayed neutron fraction data for UTR-KINKI (by MVP3.0 and JENDL-4.0)

Group (i)	Decay constant $\lambda_i [s^{-1}]$	Effective delayed neutron fraction $\beta_{eff,\,i}$
1	1.24×10^{-2}	2.423×10^{-4}
2	3.05×10^{-2}	1.608×10^{-3}
3	1.11×10^{-1}	1.439×10^{-3}
4	3.01×10^{-1}	2.900×10^{-3}
5	1.14	8.443×10^{-4}
6	3.01	3.342×10^{-4}
Total	$7.64 \times 10^{-2}*$	7.342×10^{-3} (734.2 pcm)

pcm: percent mille ($10^{-3}\%$)

*: see Ref. [6] to calculate the decay constant

1.2.8.3 Time Variation of Reactor Power

Assuming that the shape of spatial distribution of neutron flux does not change for the time variation of neutrons caused by the power change of the reactor, the "point-reactor kinetics equation" is then applied to the assumption. Here, in addition to the reactivity ρ learned in Sect. 1.2.7, the neutron generation time $\left(\Lambda = \ell/k_{eff}\right)$ is used, interpreting as "the average time taken by the generation of a new neutron triggered from another fission reaction, when neutrons produced by fission reactions are absorbed by the nuclear fuel." Instead of the prompt neutron lifetime (ℓ), the neutron generation time is used in the text, appearing in the explanation of prompt neutrons in Sect. 1.2.6. When the number of neutrons for time t is $n(t)$, the fraction of delayed neutrons β, the decay constant λ and the number of delayed neutrons per unit volume (number density) $C(t)$, the point-reactor kinetics equation can be written as follows, whereas the delayed neutrons are treated as one group instead of six groups:

$$\frac{dn(t)}{dt} = \frac{\rho(t) - \beta}{\Lambda} n(t) + \lambda C(t), \tag{1.10}$$

$$\frac{dC(t)}{dt} = \frac{\beta}{\Lambda} n(t) - \lambda C(t). \tag{1.11}$$

To understand the role of delayed neutrons, we assume that there are only prompt neutrons in the reactor. Then, for $\beta = 0$, since the second term on the right-hand side of Eq. (1.10) disappears, Eq. (1.10) can be written by using Eq. (1.9) and $\Lambda = \ell/k_{eff}$ as follows:

$$\frac{dn(t)}{dt} = \frac{k_{eff} - 1}{\ell} n(t) \text{ or } \frac{dn(t)}{dt} = \frac{\rho}{\Lambda} n(t), \tag{1.12}$$

solving Eq. (1.12) for $n(t)$, we obtain the following equation:

$$n(t) = n_0 e^{\frac{k_{eff}-1}{\ell}t} \text{ or } n(t) = n_0 e^{\frac{\rho}{\Lambda}t} \quad \because \quad n_0 = n(0). \tag{1.13}$$

From Eq. (1.13), the number of neutrons (reactor power) is found to vary exponentially. The index that expresses the speed of the exponential variation (the time required for the power to increase by a factor of e or $1/e$) is called the "reactor period."

The reactor period T in a reactor with only prompt neutrons is expressed as follows:

$$T = \frac{\ell}{k_{eff}-1} \text{ or } T = \frac{\Lambda}{\rho}. \tag{1.14}$$

Using Eq. (1.14), Eq. (1.13) can be described as follows:

$$n(t) = n_0 e^{\frac{t}{T}}. \tag{1.15}$$

1.3   Education on Operational Safety and Physical Protection of Nuclear Material

1.3.1   Overview

Nuclear-related safety, security and safeguards, called "3S" in short, are important for the peaceful use of nuclear energy. Nuclear safety here refers to the protection of humans and the environment from the effects of radiation caused by nuclear accidents. Nuclear security refers to measures taken to prevent the threat of misuse of nuclear materials or radiation sources from becoming a reality, and nuclear safeguards refer to verification activities conducted to ensure that nuclear materials are used only for peaceful purposes and are not diverted to nuclear weapons. These principles are common throughout the world, and a framework for ensuring international nuclear safety has been developed under the International Atomic Energy Agency (IAEA). The IAEA has established its Fundamental Safety Principles, which state that nuclear safety and security measures should be aimed at protecting human life and health and the environment. Based on these universal principles, it is the responsibility of each country to establish safety regulations for the peaceful use of nuclear energy. In Japan, "Act on the Regulation of Nuclear Source Material, Nuclear Fuel Material and Reactors (hereinafter, referred to as "Act on the Regulation of Nuclear Reactors")" has been established, and compliance with this act and related laws and regulations is required for the establishment and operation of nuclear facilities. The IAEA's Basic Safety Principles state that the fundamental safety goals of protecting people and the environment must be achieved without excessively restricting the operation of facilities or the conduct of activities that might pose a radiation risk. However, after the accident at the TEPCO's Fukushima Daiichi Nuclear Power Plants in 2011, the new regulatory standards have been enacted even stricter than before and are applied without exception to the extremely low-power Kindai University reactor facility. To operate and manage the facilities safely, each facility shall establish the safety program with taking into account of previsions of laws and regulations and shall comply with the program. In the same way, it is also required to establish and comply with the physical protection guideline that conforms to the specifications of nuclear security requirements.

In this section, we describe the items necessary for the safe and appropriate use of the experimental reactor facilities, recognizing that there are various compliance items under such strict regulations.

1.3.2 Radiological Safety in Nuclear Reactor Facilities

The use of radiation has brought enormous benefits to the public not only through scientific and technological research but also through industry and medicine. Also, radiation has not only such a beneficial aspect but also a harmful aspect that may cause health hazards. For example, exposure to excessive doses of radiation in a short period can cause acute damage. Even if the level of radiation does not cause acute damage, long-term exposure may cause chronic damage. In addition to these physical effects on individuals, genetic effects are also known to occur. However, it is difficult to determine whether these effects are due to radiological effects or to other factors. Therefore, the fundamental principle of radiological protection is "to reduce unnecessary radiation exposure."

On the other hand, there are quite a few cases of radiation exposure in the medical field. Furthermore, exposure to natural radiation from cosmic rays, natural radioactive materials widely distributed on the earth, and potassium-40 (K-40) and carbon-14 (C-14) in the body cannot be avoided. Since radiation cannot be felt by the five senses, in fact, the presence of radiation can only be known by measuring instruments. Through the hands-on radiation measurement practices at nuclear reactor facilities on the reactor operation training course, it is hoped that students could learn through experiences how radiation is detected, to what extent it is dangerous, and its relationship with dose limits. Here, the dose limit refers to the amount of radiation that is considered not to cause bodily harm to the extent that it can be detected at any time in one's life in light of the knowledge currently available.

The dose limits for persons engaged in radiation work are shown in Table 1.5. These dose limits are established in "Act on the Regulation of Radioisotopes, etc.," "Act on the Regulation of Nuclear Reactors," and other relevant laws and regulations, based on the recommendations of the International Commission on Radiological Protection (ICRP), which is the world's most authoritative authority on radiation protection. In addition, each plant has established specific radiological protection measures in "Programs for Prevention from Radiation Hazards" and "Operational Safety Programs." There is no need to fear radiation unnecessarily. It is important

Table 1.5 Categories of persons entering controlled area and dose limits

Classification		Effective dose limit
Radiation worker	A person who enters a controlled area to engage in the handling, management, or incidental handling of nuclear fuel materials or materials contaminated by them, and who has been designated by the director	50 mSv $(\text{year})^{-1}$ 100 mSv $(5 \text{ years})^{-1}$ Women: 5 mSv·$(3 \text{ months})^{-1}$ During the gestation period: internal exposure 1 mSv
Temporary visitor to a controlled area	Those who temporarily enter the controlled area for observation, practical training, etc	100 μSv per time

always to handle radioactive materials or radiation generators including reactors carefully and cautiously in accordance with the rules established for safe handling.

1.3.3 Precautions in Controlled Areas

The Atomic Energy Research Institute (AERI) of Kindai University has a nuclear reactor facility and a radioisotope facility, and the boundary of the plant including these facilities is designated as an "environmental monitoring area" by Act on the Regulation of Nuclear Reactors. The environmental monitoring area is shown in Fig. 1.14. In the environmental monitoring area, entry is restricted, and the dose limit outside the boundary of the area (1 mSv $y^{-1} = 0.11$ μSv h^{-1}) is also set. Students participating in the reactor operation training are obliged to carry the entry permit in a visible position at all times in the environmental monitoring area.

The radiation-controlled area of the UTR-KINKI facility is also shown in Fig. 1.14. In this controlled area, it is necessary to take care not to be exposed to radiation unnecessarily and to take care not to contaminate laboratory equipment, articles, and clothes. As shown in Table 1.5, those who enter the controlled area are classified into radiation workers and temporary visitors. Trainees who are temporary entrants shall limit their exposure to 100 μSv or less per visit and shall observe the following rules in the controlled area:

(1) When entering the controlled area, wear special work clothes, slippers and an electronic personal dosimeter.
(2) Eating, drinking, smoking and makeup are not allowed in the controlled area.
(3) Keep the facilities in order at all times and do not bring into the controlled area any items other than the minimum necessary for the experiment.
(4) When leaving the controlled area, take off your laboratory clothes and use a hand–foot cloth monitor or a surface contamination survey meter to check for surface contamination on your hands, feet, clothes and items you are taking with you, and confirm that there is no contamination before leaving.

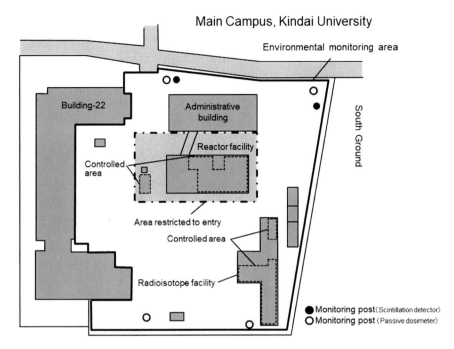

Fig. 1.14 Map of environmental monitoring area around AERI in Kindai university

Radiation control involves measurements of radiation dose, concentration of radioactive materials and surface contamination in a facility. The objectives of radiation control are to protect human body from radiation, to create and maintain a safe working environment for workers and to confirm that the working environment is in accordance with the design standards. It is important to control the internal dose as well as the external exposure dose, and the measurement of radioactive concentration is classified into air contamination and surface contamination. Ambient dose and personal dose measurements should be carried out in the work space, controlled area and surrounding monitoring area for each work environment, and it is necessary to control the dose to keep it lower than the dose limit. In general, the concentration of radioactive materials in the atmosphere is measured by using radioactive gas monitors and radioactive aerosol monitors in the controlled area and at the outlet, and the ambient dose measurement is continuously monitored by area monitors or monitoring posts. In addition, surface contamination is measured at least once a week.

The maximum thermal power of UTR-KINKI is a sufficiently small power of 1 W, which is sufficiently small to keep a safe operation without any cooling system and produce few fission and corrosion products. As a result, no gaseous and liquid waste needs to be treated before discharge are generated. The concentration of radioactive materials in the effluent is measured before discharge, and it is confirmed that the concentration is below the limit.

1.3.4   Handling of Nuclear Fuel Materials

At UTR-KINKI, the highly-enriched uranium (HEU) fuel plates are loaded, and a Pu-Be neutron source is used for a reactor start-up and various neutronics experiments. The structure of the fuel plate is shown in Fig. 1.5. The fuel meat is made of U-Al alloy, which is coated with Al to make a single fuel plate. Therefore, the nuclear fuel material is hermetically sealed and will not leak and contaminate surroundings under normal operating conditions.

In handling facilities of nuclear fuel materials, it is generally required to comply with both the safety regulations on the handling and the protection regulations on the protection of nuclear materials. Safety in handling facilities includes criticality safety control to prevent nuclear fuel materials from critical accidents, and safety control mainly for radiation dose control in radiological protection. Physical protection is the protection of nuclear materials and facilities from theft and sabotage of nuclear materials during use, storage and transport. Since the 1960s, terrorism has been occurring frequently in the USA and on an international scale, and there have been fears of theft and sabotage of nuclear materials, and attacks on nuclear facilities. In 1987, the Convention on the Physical Protection of Nuclear Material came into effect in the international framework. Japan is obliged to take measures for the physical protection of nuclear materials by the Nuclear Reactor Regulation Law and other laws based on this international agreement. Therefore, it is necessary to recognize that there are various rules and restrictions on the handling of nuclear fuel under these protective measures. The protective measures in the UTR-KINKI facility include the establishment of protected areas, etc., monitoring and patrolling, installation of protective equipment and devices, and access control to the facility (Fig. 1.14).

1.3.4.1   Entry into Restricted Areas

A restricted-entry area is set for the protection of specific nuclear fuel materials (protected area). When a person enters a protected area or a restricted access area, his or her identity is checked by a security guard or a laboratory personnel. An official identification card with a photograph, such as a driver's license (a student ID card with a photograph is also acceptable for identification of students), is required for to enter into the restricted area. If you do not have an ID, you will not be allowed to enter the restricted area for any reason (you will not be able to participate in any on-site experiments or training). Those who have completed the identification will be lent an entry/exit permit. The restricted area is patrolled by security guards, and you are required to keep your entry permit visible at all times in the restricted area.

Gate-type metal detectors are installed at the boundaries of restricted areas to prevent the entry of items that may be used for such sabotage or other illegal activities. Visitors are required to remove any metal objects from their bodies before passing through the gate. If they are detected, they will be checked by a portable metal detector. Security guards may also conduct a body check if they deem it necessary.

In principle, students are not allowed to bring in any items other than those necessary for practical training and experiments.

1.3.4.2   Protected Area

All nuclear materials are stored and used in the protected area. The protected area is surrounded by a restricted access area, with stricter restrictions imposed as the area approaches the location where nuclear materials are stored and used. The protected area and the restricted area are controlled so that no one other than authorized persons may enter them.

Persons who enter the protected area at any time are required to have a certificate issued to them after confirming the necessity of entry, and several actions are taken to prevent persons without such a certificate from entering the area. A person who enters the protected area for temporary work for the reactor operation training or other experiments (hereinafter referred to as a temporary entrant) is not allowed to enter the area without a certificate. Hereinafter referred to as "temporary entrants" are required to have an entry permit issued after confirming their status and the necessity of entry. For students who enter for the reactor operation training or other experiments, this check is conducted at the same time as the identification check when entering the restricted area. When temporary visitors enter the restricted area, they must be accompanied by a permanent visitor of AERI. The full-time entrant with permission must supervise temporary visitors as necessary for the protection of nuclear materials, and temporary visitors must follow the instructions of the permanent visitor.

1.3.4.3   Precautions for Handling Nuclear Fuel

In addition to the U-235 fuel, UTR-KINKI has other experimental nuclear fuels and a Pu-Be neutron source, and prior permission is then required for handling any of them. Other experimental nuclear fuels should be used at the "nuclear fuel material use site" or the "nuclear fuel material handling site." The following items must be observed when using them:

(1) To prevent criticality from being reached outside the reactor.
(2) The type, quantity, etc., shall be checked and recorded each time when fuels are used.
(3) When inserting fuels into the reactor, fuels should be handled carefully with the consideration of the effect on the reactivity.

1.3.4.4   Procedures for Using Pu-Be Neutron Source

As described in Sect. 1.3.4.3, prior permission is required to use the Pu-Be neutron source. The necessary procedures are carried out by the AERI staff in charge of reactor experiments and practical training programs.

The purpose, method, date and time of use, place of use, and working conditions (distance and time) must be clarified before use. Therefore, you can only use the facility within the scope of the permission granted in the application for use. Also, even if you suddenly want to conduct experiments under conditions other than those for which you applied, you cannot carry out experiments if experiments are not recognized as the scope of the permission. In this case, the user must strictly observe the precautions stated in the permit.

1.3.4.5 Handling of Reactor Fuel Elements

Handling of reactor fuel elements in the reactor room shall be performed in compliance with the following items:

(1) The work shall be carried out in accordance with the "Work Plan for Changing the Loaded Fuel" and the "Work Plan Approval Document."
(2) The work shall be carried out under the supervision of the person in charge of the work.
(3) The reactor fuel plates should be handled in the air so as not to reach criticality under any circumstances, and the number of fuel plates to be handled simultaneously should be limited to six fuel plates at most.
(4) The reactor fuel assembly shall not be damaged by a strong impact or other events.
(5) A person shall be assigned to monitor the reactor fuel assembly while handling.
(6) After completion of the work, the person responsible for the work shall describe the results of the work in a "Work Report" and submit it.

Although the students who participate in the reactor operation training and other experiments do not actually handle fuel elements, the students may remove fuel assemblies from the reactor, insert fuel assemblies into the core, change their positions in the reactor, or reload fuel assemblies during the approach to criticality experiment.

1.3.5 Actions of Emergency

In nuclear facilities, it is generally required to ensure safety even in an emergency. Emergencies in this context include the following events:

- Fire
- Earthquake
- Occurrence of life-threatening events (accidents, sudden illness, etc.)
- Discovery of suspicious persons
- Abnormalities in reactor equipment

- Students (temporary visitors) staying in the reactor facility only without the AERI staff.

Since it is necessary to get permission from the AERI staff to enter the reactor facility at any time, a situation in which only students are in the facility is considered an emergency. Regardless of the above, if an unknown event occurs, it is mostly important to inform the AERI staff.

(1)   Fire

To prevent the occurrence of fire in the reactor facility, combustible materials are restricted from being brought into the reactor room and the reactor control room. In the unlikely event of fire, the following shall apply:

- Shout out to the people around you, "Fire!" to those around them.
- Follow the instructions of the AERI staff.
- If the reactor is in operation, contact the person in charge of the operation to stop the reactor immediately.

(2)   Earthquake

Secure your own safety first. If the earthquake is large enough, the reactor will automatically shut down. In the event of an earthquake, follow the instructions of the staff and evacuate.

(3)   Life threatening (accident, sudden illness)

Notify those around you in a loud voice and follow the instructions of the staff. If necessary, rescue the injured person, give first aid and call the management office.

(4)   Discovery of suspicious persons

Anyone not wearing an access permit is considered a "suspicious person." If you know someone who is not wearing an entry permit, instruct them to wear it, and if they are not an acquaintance, contact the management office. Do not make inadvertent contact with suspicious persons.

(5)   Abnormalities in reactor equipment

If the red light is on the alarm panel or the emergency stop panel (scram) of the control console, notify nearby the staff and wait for instructions. Do not reset the alarm without instructions to determine the cause of the alarm/scram.

(6)   Students alone in the reactor facility

Temporary visitors are not allowed to remain in the reactor facility. They should leave with the staff when they leave. If the staff tries to leave without the students (temporary visitors), they will be held back.

In the unlikely event that you are left behind in the reactor facility, immediately contact the control room.

1.4   Reactor Operation

1.4.1   Purpose

UTR-KINKI is designed and produced for the purpose of education and training at universities, and one of its main characteristics is that students can operate the reactor themselves for practical training. Since it is an extremely safe reactor, even if you make a mistake in operation, there will be no radiation hazard or severe accident that could affect the surrounding environment. Here, you are encouraged to actively engage in reactor operation without fear of making mistakes. However, regardless of the small thermal power and safety of the reactor, UTR-KINKI is a nuclear facility that is strictly managed and operated under laws and regulations. From the start-up to shut down, you must ensure that not only the reactor operation but also instrument checking and data recording are carried out in accordance with prescribed procedures and rules. Although your supervisor will give you instructions for these tasks at each stage of operation, it is important that you do not blindly follow the instructions, but that you understand the technical and legal background of each task as you perform it.

In practical training, reactor operation may be conducted as a stand-alone training program, or the reactor operation may be combined with the various training programs described in Chap. 2 onward. The practical training involving reactor operation should provide an opportunity for students to put into practice the knowledge of nuclear reactors and radiation that they have learned in the classroom so far, since students will operate the reactor according to the procedures described in this section.

The following subsection describes the control console of UTR-KINKI, followed by a series of operation procedures according to the general sequence of operation.

1.4.2   Reactor Control Console

The control console of UTR-KINKI consists of several panels that are necessary for the operator to control the reactor and to monitor the reactor status, as shown in Fig. 1.15. The panels are called A, B, C and D from left to right, and 1, 2, 3, … from top to bottom in order. For example, if you are told "B-3," you are referring to the picoammeter. The main functions of the panels related to the reactor operation are described in this subsection.

1.4.2.1   Picoammeter (Lin-N Meter) and Power Recorder (Chart Recorder)

The current signal obtained from the CIC that monitors the reactor power is directly measured by a picoammeter and displayed on the linear power meter (Lin-N meter).

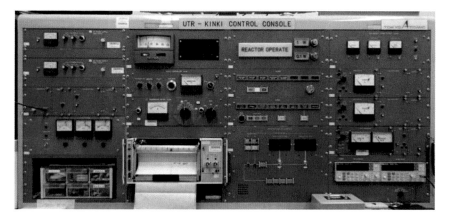

Fig. 1.15 Control console of UTR-KINKI (© AERI, Kindai University. All rights reserved)

The output signal from the linear power meter is continuously recorded on a chart paper by a power recorder (chart recorder). When the thermal power of the reactor is 1 W (rated thermal power), the linear power meter indicates 5.72×10^{-8} A. From this relationship between the reactor power and the reading of the linear power meter, the reactor power is monitored and the reactor is operated.

1.4.2.2 Period Meter

The signal from the CIC of the intermediate channel is used to indicate the reactor period (the time required to change the power by a factor of e). The shorter the reactor period is, the more rapidly the reactor power increases, which is undesirable for reactor control. If the period is shorter than 30 s, a warning sound is made, and the reactor must be operated so that the period does not become shorter than 30 s. If the period becomes shorter than 10 s, an alarm is triggered, and if the period becomes shorter than 5 s, the reactor is immediately shut down.

1.4.2.3 Alarm

The alarm is provided to alert the operator when the reactor operating condition may deviate from the normal range. The alarm circuit of the UTR-KINKI is shown in Fig. 1.16. The operating conditions of the alarm are as follows:

(1) PERIOD UNDER 10 s
    Condition: when a period is shorter than 10 s.
(2) POWER RECORDER OFF SCALE
    Condition: when the indication of the chart recorder swings out of the display range.

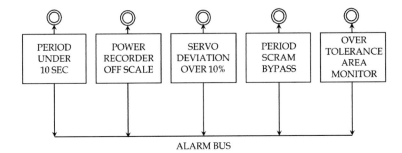

Fig. 1.16 Alarm circuit of UTR-KINKI

(3) SERVO DEVIATION OVER 10%

Condition: when the relative deviation of the power from the demand level for automatic operation is greater than or equal to ±10%.

(4) PERIOD SCRAM BYPASS

Condition: when the scram signal due to a short period is bypassed by pressing the scram bypass button.

(5) OVER TOLERANCE AREA MONITOR

Condition: when the readings of the γ ray area monitors installed at the top of the reactor, at the side of the reactor (under the stairs), and at the west wall of the reactor room exceed the set values. The set values are 50 μ Sv h^{-1} for the top of the reactor and 20 μ Sv h^{-1} for the side of the reactor and the west wall of the reactor room.

1.4.2.4 Scram

When the reactor must be shut down immediately, a scram signal is triggered by the safety protection system of the reactor. When a scram signal is generated, the electromagnetic current of the electromagnetic clutch of the control rod drive mechanism is shut off and the three safety rods (SR#1, SR#2, and SSR) drop into the reactor core, resulting in an emergency shut down (scram) of the reactor. The time required to insert the control rods is within 0.5 s. The scram circuit of UTR-KINKI is shown in Fig. 1.17. The operating conditions of the scram are as follows:

(1) EARTHQUAKE

Condition: when the seismic detector detects an acceleration of 100 gals or more.

(2) PERIOD UNDER 5 s

Condition: when the reactor period is shorter than 5 s.

(3) LOW LEVEL SHIELD TANK

Condition: when the water level in the biological shielding tank drops below 160 cm.

(4) TOP CLOSURES OPEN

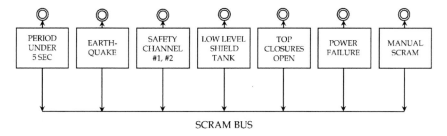

SCRAM BUS

Fig. 1.17 Scram circuit of UTR-KINKI

Condition: when the top shielding closures are opened.

(5) POWER FAILURE

Condition: when electric power supply to the control console is lost.

(6) SAFETY CHANNEL #1 and #2

Condition: when the readings of the UICs installed in the reactor exceed 150% of rated thermal power. There are two independent channels.

(7) MANUAL SCRAM

Condition: when the operator judges that the reactor needs to be shut down immediately, the operator can press the manual scram button to shut down the reactor. The manual scram button is also used for the normal shutdown of UTR-KINKI.

1.4.3 Pre-Startup Inspection

Before starting reactor operation, pre-startup inspections are conducted to confirm that the equipment for reactor control and protection is working properly. At UTR-KINKI, pre-startup inspections are conducted once a day, prior to the first operation of the day. Students may perform all inspection items in the reactor operation training, but if time is limited, UTR-KINKI staff may complete the inspections in advance and students may only inspect representative items. The list of inspection items are shown in Tables 1.6, 1.7 and 1.8.

1.4.4 Reactor Start-Up

The reactor cannot be started up and the control rods cannot be withdrawn unless the following procedures are performed in the order of steps (1) to (4):

(1) Insertion of neutron source
(2) Withdrawal of SR#1
(3) Withdrawal of SR#2

Table 1.6 Inspection items for pre-startup inspection

Inspection Item	Panel	Procedure
Operation permit	–	Check that an operation permit and a weekly operation plan are displayed in the designated area
Power supply	–	Check that the power supply required for reactor operation is normal
Exhauster and gas monitor	–	Check that the exhauster and the gas monitor are working normally
Weather condition	–	Check that the weather condition is not unsuitable for reactor operation
Control console power supply	–	Check that the push-button switchgears (CB1 and CB2) of the main power supply of the control console are turned ON
	C-1	Turn the key switch ON according to the permitted power (0.1 W or 1 W). This time is recorded as the operation start time
Moderator temperature	B-1	Check and record the moderator temperature
High voltage power supply	A-1 A-2	Check that + 1200 V is supplied to two CICs
Startup trip	A-3	Confirm that the "Low count rate" lamp turns ON and the FC count rate is less than 10 cps
Digital rate meter	D-5	Check the display settings of the digital rate meter, which displays the FC count rate
Log-N & period meter	D-4	Input a simulated signal and confirm that the operating status and the indication of the Log-N meter and the period meter are normal
Safety amp	D-1	Confirm that the electromagnet ammeter of the control rod drive mechanism of SR#1, SR#2 and SSR indicate the appropriate range
Picoammeter	B-3	Set the range selection switch to the 1×10^{-11} A position Set the mode selection switch to ZERO and adjust the picoammeter indication to 0
Recorder	B-4	Lower the pen of the chart recorder and start recording on the chart paper
Servo controller	B-2	Check the settings of the servo controller used for automatic operation MANUAL/AUTO switch: MANUAL Reset rate: 100 Proportional band: 30 Rate time: 4.9 Power demand: 572

(continued)

Table 1.6 (continued)

Inspection Item	Panel	Procedure
Reactor startup control	C-4	Confirm that the "DOWN" lamps of the SR#1 and SR#2 are ON Confirm that the position indicators of SSR and RR are 0% Confirm that the "AUXILIARY INTERLOCKS" and "SCRAM BUS CONTINUITY" lamps are ON Confirm that the "MINIMUM COUNT-RATE" and "ROD WITHDRAWAL PERMIT" lamps are OFF
Alarm test: see Table 1.6		
Scram test: see Table 1.8		
Operation indicator	C-1	Confirm that the "Reactor operate" lamp is ON Confirm that the "Reactor in operation" is displayed on the electronic bulletin boards at three locations in the reactor building
Area monitor for power failure	–	Turn on the power of the area monitor recorder to be used during a power failure

Table 1.7 Alarm test

Confirm that the buzzer sounds at each item and the indicator of the corresponding item on the alarm panel (C-2) turns on

PERIOD UNDER 10 s	D-4	Input a simulated signal to the period meter and confirm that an alarm is triggered when the reactor period is 10 s or less
POWER RECORDER OFF SCALE	B-3	Confirm that an alarm is triggered when the indication of the chart recorder swings out of the display range
SERVO DEVIATION OVER 10%	B-2	Confirm that an alarm is triggered when the operation mode is switched from MANUAL to AUTO
PERIOD SCRAM BYPASS	C-3	Confirm that an alarm is triggered when the SCRAM BYPASS button is pressed
OVER TOLERANCE AREA MONITOR	A-4	Input a simulated signal and confirm that an alarm is triggered when the indication exceeds the set value

(4) Withdrawal of SSR and RR.

For example, if you try to operate step (2) ignoring step (1), the control console will not accept your operation, and, as a result, the reactor will not be operated. This is because a safety mechanism called an interlock, which does not accept the next operation until a certain condition is satisfied, works as a mechanism to prevent

Table 1.8 Scram test

Confirm that the electromagnet ammeter (D-1) of the control rod drive mechanism indicates 0 for each item, and that the indicator lamp of the corresponding item on the scram panel (C-3) turns ON, and that the "Reactor Operate" lamp turns OFF

EARTHQUAKE	-	Confirm that a scram is triggered when the contacts of the seismic detector are disconnected
PERIOD UNDER 5 s	D-4	Input a simulated signal to the period meter and confirm that a scram is triggered when the reactor period is 5 s or less
SAFETY CHANNEL #1 and #2	D-2 D-3	Input a simulated signal to the power meter of the safety channel and confirm that a scram is triggered when the indication is 150% or more
LOW LEVEL SHIELD TANK	–	Confirm that a scram is triggered when the water pressure is reduced
TOP CLOSURES OPEN	–	Confirm that a scram is triggered when the top shielding closures are lifted
POWER FAILURE	D-1	Confirm that a scram is trigged when the power switch on the back of the control console is turned off
MANUAL SCRAM	C-3	Confirm that a scram is triggered when the manual scram button is pressed

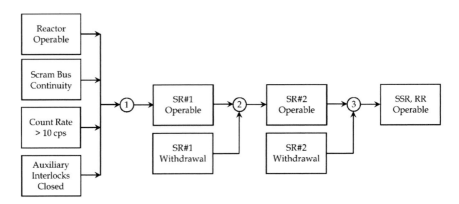

Fig. 1.18 Control rod withdrawal interlock circuit of UTR-KINKI

the operator from making a mistake. The control rod withdrawal interlock circuit of UTR-KINKI is shown in Fig. 1.18.

1.4.4.1 Neutron Source Insertion

Before the reactor is started up, of course, no fission chain reaction has occurred in the reactor. Neutrons are needed to start fission chain reactions, but the count rate from the FC is almost 0 cps before the reactor is started up. At this time, the "Minimum Count Rate" lamp on the Reactor Startup & Control panel (C-4) is off, indicating that the minimum neutron count rate required to start the reactor has not yet been obtained. This is one of the interlocks, and if the count rate is below 10 cps and this lamp does not turn on, the next operation cannot be performed.

To obtain a sufficient neutron count rate, a startup neutron source is first inserted into the reactor to supply the reactor with neutrons to trigger fission reactions. The neutron source can be inserted manually into the neutron source insertion hole in the top shielding closure of the reactor. When the neutron source is inserted, the reading of the picoammeter rises sharply and an alarm is triggered as the reactor period drops below 10 s, and a scram is triggered when the reactor period drops below 5 s. This is, however, due to an apparent power increase in the process of inserting the neutron source, and not a state in which the reactor reactivity is positive and the exponential power increase continues. Therefore, once the neutron source is inserted into a predetermined position, the reading of the picoammeter becomes constant and the reading of the period meter becomes infinite (∞). Once this condition is confirmed, the alarm and scram are reset. The insertion of the neutron source is recorded in the operation log and on the chart paper with the insertion time.

After the neutron source is inserted, the "Minimum Count Rate" lamp (C-4) turns on as the FC count rate exceeds 10 cps, and the next operation, SR#1 withdrawal is ready.

The startup neutron source of UTR-KINKI is a Pu-Be neutron source (1.4×10^6 $n\ s^{-1}$). This is a sealed source consisting of a mixture of Pu and Be encapsulated in a stainless steel capsule. Neutrons are generated by the following nuclear reactions of α-rays (He-4 nucleus) emitted by the α-decay of Pu with Be-9 nucleus.

$$^9\mathrm{Be} + {}^4\mathrm{He} \rightarrow {}^{12}\mathrm{C} + {}^1n.$$

1.4.4.2   Withdrawal of Safety Rods

The Reactor Startup & Control panel (C-4) has two lamps to indicate the positions of SRs #1 and #2 (UP or DOWN) and two buttons to start withdrawing them from the core. Before starting operation, the DOWN lamps should be turned on for both safety rods, indicating that the two safety rods are inserted into the lower limit positions. Due to the interlock, SRs #1 and #2 cannot be withdrawn at the same time, and SR#2 can only be withdrawn after SR#1 has been withdrawn to the upper limit position.

When the SR#1 withdrawal button is pressed, the control rod drive mechanism is activated to start the withdrawal of SR#1. In UTR-KINKI, the start time of the SR#1 withdrawal is defined as the reactor start-up time. When the SR#1 withdrawal is started, use the broadcasting facility in the control room to announce that the reactor has been started up, and record the start of the SR#1 withdrawal in the operation log with the withdrawal start time.

A short time after the button is pressed, the DOWN lamp turns off, indicating that the SR#1 is positioned between the upper and lower limits. The SR#1 is then withdrawn slowly over about 180 s, during which time observe that the reading of the picoammeter rises. Operate the range selection switch of the picoammeter appropriately, and be careful that the reading of the chart recorder does not exceed the display range (exceeding the display range will trigger the "POWER RECORDER OFF SCALE" alarm).

When the SR#1 reaches the upper limit position, a chime sounds and the UP lamp turns on. In this condition, the SR#2 can be withdrawn, so press the button to start withdrawing from the core. As in the case of SR#1, the DOWN lamp of the SR#2 turns off after a while, and the SR#2 is withdrawn to the upper limit position in about 180 s, a chime sounds and the UP lamp turns on. Once the UP lamps are on for both SRs #1 and #2, the SSR and RR can then be operated.

1.4.4.3 Withdrawal of Shim Safety Rod and Regulating Rod, and Extraction of Neutron Source

When SR#2 is withdrawn to the upper limit position, the Sim Safety Rod (SSR) and Regulating Rod (RR) can be operated. Unlike the SRs, the SSR and RR can be operated up and down with an operating switch and held at any position. The withdrawal ratio (%) of the SSR and RR is digitally displayed. For example, the withdrawal ratio is displayed as 0% when the control rod is at the lower limit position, and 100% when it is at the upper limit position.

The SSR is a control rod with the same large negative reactivity as the SR and is a coarse control rod that moves up and down slowly. Meanwhile, the RR is a fine control rod with reactivity worth as small as $0.1\% \Delta k / k$ and can be operated up and down faster than the SRs and SSR.

When the SSR and RR are withdrawn, the reading of the picoammeter and the chart recorder will increase. At this time, the reading of the period meter should not be shorter than 30 s. If the period falls below 30 s, a warning (not an alarm) will sound, and the SSR and RR must be inserted into the core immediately so that period should be longer than 30 s.

When the reactor power increases and the reading of the picoammeter reaches about 4×10^{-10} A, the neutron source is removed from the core, and the removal is recorded in the operation log and on the chart paper. Then, it can be observed that the reading of the chart recorder decreases rapidly. This means that the reactor is still at a subcritical state. Here, if the reactor is left at the subcritical state, the reactor power will decrease and return to the status before startup. Therefore, the SSR must be withdrawn further until the indication of the chart recorder begins to rise. When the indication of the chart recorder begins to rise without the neutron source, the reactor is at a supercritical state, where the reactor power is increasing solely by the fission chain reactions in the reactor.

1.4.5 Reactor Criticality and Power Change Operation

1.4.5.1 Critical State Operation at 0.01 W

When the indication of the picoammeter approaches 5.72×10^{-10} A, which corresponds to 0.01 W of thermal power, operate the SSR and RR to bring the reactor from supercritical state to critical state. When approaching criticality, carefully monitor the increase or decrease of the chart recorder indication (the line drawn on the paper), and insert the control rods if the power is increasing, and withdraw the control rods if the power is decreasing.

Here, observe carefully how the reactor power changes when the control rod is operated to add positive or negative reactivity to the reactor. You will see that the reactor power changes rapidly immediately after the control rod is operated, but as time passes, the change in power becomes more gradual. This phenomenon is attributed to the fact that immediately after the reactivity is added, the influence of prompt neutrons is dominant, but the prompt neutron component disappears in a short time, and the change in reactor power becomes dominated by the influence of delayed neutrons. Therefore, when approaching criticality, it is necessary to observe the change in reactor power for a while after the control rod is operated and to perform the next operation only after the change has stabilized.

When the indication of the chart recorder becomes constant within a range of statistical variation after the repeated control rod operations, the reactor is critical. Record the power and the time when criticality is reached on the chart paper and operation log, and broadcast that the reactor has reached criticality.

1.4.5.2 Automatic Operation and Checking

The UTR-KINKI can be operated in the automatic operation mode. When instructed by the supervisor to start automatic operation, the switch on the SERVO CONTROLLER panel (B-2) should be switched from MANUAL to AUTO. When switching from MANUAL to AUTO, record the time in the operation log. This operation will allow the regulating rod (RR) to automatically move up and down to maintain the reactor in a critical state.

Regardless of how much power the reactor is operated, the regulations require that the reactor be operated at the critical state of 0.01 W at the first operation of the day, and that the equipment for reactor control and protection be checked to confirm that the reactor is operated normally. A few minutes after starting automatic operation, the following items should be checked and recorded in the operation log.

- Time of checking
- Reactor power (W)
- Operation status (automatic/manual)
- Position of SSR (%)
- Position of RR (%)

Table 1.9 Example of reactor operation condition

Indicated value of picoammeter (A)	Reactor power* (W)	Position of SSR (%)	Position of RR (%)
2.56×10^{-9}	0.5	92.0	0

* The reactor power is calibrated as 1 W when the signal of the picoammeter is 5.72×10^{-8} A

- Picoammeter reading (A)
- Safety channel readings (UIC) (%) (#1 and #2)
- Log-N meter reading (CIC) (A)
- Digital rate meter reading (CPS)
- Moderator temperature (°C)
- γ-ray area monitor readings (μSv·h^{-1}) (#1 and #2).

1.4.5.3   Power Increase and Full Power Operation at 1 W

After the checking at 0.01 W is completed, switch the operation mode from AUTO to MANUAL, and operate SSR and RR to make the reactor supercritical again to increase the reactor power. At this time, record the switching from AUTO to MANUAL and the start of the power increase with time in the operation log. When the indication of the picoammeter approaches 5.72×10^{-8} A, which corresponds to 1 W of thermal power, operate the SSR and RR again to bring the reactor closer to the critical state. The operation procedure until the reactor reaches criticality and then shifts to automatic operation is the same as that used when the criticality was achieved at 0.01 W.

Once the reactor is in automatic operation at the rated thermal power of 1 W, check the items shown in Sect. 1.4.5.2 again to confirm that the reactor is operated normally. If the automatic operation continues for a long time, repeat this check every hour.

1.4.5.4   Power Change

Using the same operating procedure as in the previous cases, decrease or increase the reactor power to the power indicated by the supervisor to bring the reactor critical. Every time the reactor reaches criticality, record the picoammeter reading, reactor power, and SSR and RR positions as shown in Table 1.9. Note that the reactor can be brought to criticality at any power.

1.4.5.5   Measurement of Air Dose Rate

When operating UTR-KINKI, the regulations require that air dose rates be measured at designated locations around the reactor when it is operated at its maximum power for the day. During the automatic operation at 1 W, use the ionization chamber and neutron rem counter provided in the reactor room to measure air dose rate for γ-rays and neutrons.

A detailed description of the measurement is given in Sect. 3.3 of Chap. 3, "Measurement of Neutron and Gamma Radiation Air Dose Rates."

1.4.5.6 Reactor Shutdown

When the scheduled operation is completed, the reactor is shut down. In UTR-KINKI, the SSR and RR can be inserted at any position with the operating switch, but the two SRs only have a button to start withdrawing. Therefore, the reactor is shutdown by pressing the manual scram button. When the manual scram button is pressed, SR#1, SR#2 and SSR are rapidly inserted into the reactor.

If the reactor is in automatic operation, switch to manual operation and press the manual scram button to shut down the reactor. The time of switching from AUTO to MANUAL and the time of reactor shutdown should be recorded in the operation log. Confirm that the three control rods are fully inserted by checking that the two SRs are inserted and both "DOWN" lamps turn on, and that the SSR position indicator is at 0%. The RR should be inserted manually to 0% with the operating switch.

After the reactor is shut down, carefully observe the change in the picoammeter reading and the counting rate from the FC. You may want to use a scaler provided in the control room to record the integrated counts from the FC every 10 s. Then you can see that the reactor power drops significantly immediately after the control rods are inserted, and gradually declines. The reactor power should decrease in accordance with the half-life of the group of delayed neutron precursors with long half-life.

When the reactor power is sufficiently low and the "LOW COUNT RATE" lamp (A-3) turns on, meaning that the FC count rate is less than 10 cps, post-shutdown check is performed for the specified items. When the check is completed, the reactor operation is finished.

1.4.6 Discussion

(1) Consider the reason why a neutron source is first inserted to supply neutrons to the reactor when the reactor is started up. If the control rods were withdrawn without inserting a neutron source, what problems would result?

(2) From Table 1.9, it can be seen that the positions of the SSR and RR are almost the same at any reactor power when the reactor is critical. Consider the reason.

(3) Suppose the reactor power is increased beyond the rated thermal power of 1 W and the reactor is critical at higher powers such as 100 W or 1 kW. How would the position of the control rod change in the critical state at such a high power, taking into account the temperature change of the fuel and moderator?

(4) From the results of the air dose rate measurements, it can be seen that the dose rates of γ-rays and neutrons vary considerably depending on the location of the measurements. Consider the reason.

(5) Can the change in the count rate after the reactor shutdown be explained by the half-life of the group of delayed neutron precursors with long half-lives? Refer to Table 1.2 for discussion.

References

1. Tajima I (2005) A way of life—Koichi Seko. Kinki University, p 73. (in Japanese)
2. Miki R (1995) 30-odd years with UTR. Annu report of atomic energy research institute, vol 32. Kindai University, pp 1–5. (in Japanese)
3. Atomic Energy Research Institute, Kindai University (2008) Application for change of reactor installation permit (permitted on 11 May 2008). (in Japanese)
4. Nagaya Y, Okumura K, Mori T (2015) Recent development of JAEA's Monte Carlo code MVP for reactor physics applications. Ann Nucl Energy 82:85–89
5. Shibata K, Iwamoto O, Nakagawa T et al (2011) JENDL-4.0: a new library for nuclear science and engineering. J Nucl Sci Technol 48:1–30
6. Atomic Energy Society of Japan, Reactor Physics Division ed (2008) Reactor physics (textbook of reactor physics: intermediate edition). Atomic Energy Society of Japan, Tokyo, Japan. (in Japanese) https://rpg.jaea.go.jp/else/rpd/others/study/text_aesj.html. Accessed 1 July 2022

Chapter 2
Reactor Physics Experiments

Abstract A correct comprehension of fission chain reactions in a reactor core is
a significant clue for carrying out reactor physics experiments, including approach
to criticality, control rod calibration and subcriticality experiments as described in
this chapter. Based on an easy understanding of neutron characteristics on a reactor
core, such as fission chain reactions, approach to criticality experiment is the start
of the first touch of reactor physics itself. Before explaining about the approach
to criticality experiment, neutron multiplication is briefly described by using two
theoretical concepts with and without a neutron source. In terms of control rod calibration experiment, measurement methodologies are provided for excess reactivity
and control rod worth by the positive period and the rod drop methods, respectively.
Noteworthy is that subcriticality measurement methodologies are uniquely introduced, including the neutron source multiplication method, the source jerk method,
the inverse kinetics method and the reactor noise analysis method. Also, the methodologies could be helpful to understand the magnitude of negative reactivity in the
core: to offer a kindly response to "how much margin (i.e., subcriticality) is left in
the core from the critical state?".

Keywords Approach to criticality · Control rod calibration · Subcriticality

2.1 Approach to Criticality Experiment

2.1.1 Purpose

In fission chain reactions, fission reactions are induced by the U-235 absorption reactions with neutrons, and two or three neutrons generated at this time are moderated
to be thermal neutrons, and one or more of the neutrons could contribute to fission
reactions in a next generation. To reduce the ratio of wasted neutrons, it is necessary
to reduce the number of the neutrons that are absorbed into U-238 and other structural materials of the reactor or leaked out of the core. Thus, we would like you to
learn how to appropriately control the absorption and leakage of neutrons through
the approach to criticality experiment, in addition to fission reactions in the system.

© The Author(s) 2023
G. Wakabayashi et al., *Introduction to Nuclear Reactor Experiments*,
https://doi.org/10.1007/978-981-19-6589-0_2

When a new reactor core is constructed in a reactor building, fuel is loaded into the core little by little. Then, to confirm the birth of the new reactor, a first experiment is always conducted by measuring the amount of the fuel that is just enough to induce fission chain reactions, i.e., the amount of the fuel that is sufficient to reach a critical state. The experiment is called the approach to criticality experiment. The main purpose of the experiment is to understand how the critical state is reached in UTR-KINKI.

2.1.2 Principle of Measurement

2.1.2.1 Neutron Multiplication

(1) Balance of neutrons

Balance of neutrons in a nuclear reactor was explained in Sect. 1.2.5, as can be summarized as follows: "Neutrons are generated by fission reactions, move around the reactor repeatedly by scattering reactions, and eventually leak out of the reactor or are absorbed by structure materials in the reactor."

The balance of neutrons is assumed to be in a nuclear reactor, and then, let the balance in a system with a certain volume, and focus on the change in the number of neutrons in unit volume and unit time. When the increase or decrease of the number of neutrons is considered as a positive or negative effect, the rate of change can be finally expressed by using the effects of increase and decrease as follows:

$$\text{"Rate of change"} = (\text{Effect of increase}) - (\text{Effect of decrease})$$
$$= (\text{"Production rate"} + \text{"Inflow rate"})$$
$$- (\text{"Absorption rate"} + \text{"Outflow rate"})$$

We transform "Rate of change" mentioned above as follows:

$$\text{"Rate of change"} = \text{"Production rate"} - \text{"Absorption rate"}$$
$$- (\text{"Outflow rate"} - \text{"Inflow rate"})$$

If the third term on the right-hand side in the above "equation" is "net outflow rate (leakage rate)," the "Rate of change" can be rewritten as follows (We call this the "equation of continuity," but we skip the details):

$$\text{"Rate of change"} = \text{"Production rate"} - \text{"Absorption rate"} - \text{"Leakage rate"}$$

Let's imagine which reaction in the behavior of neutrons in a nuclear reactor corresponds to three terms that constitute the above "Rate of change." Here, we consider that the "rate (probability) is proportional to the number of neutrons. The "Production rate" is the number of neutrons generated by fission reactions, the "Absorption rate" is that by absorption reactions, and the "Leakage rate" is that by leaked outside. Furthermore, when the three terms are divided into positive and negative groups again, the overall balance between two groups, that is, the ratio of "increase due to positive factors" and "decrease due to negative factors" is not always unity. When the coefficient that adjusts this balance (adjustment factor) is k, k can be expressed as follows:

$k =$ "Increase due to positive factors"/"Decrease due to negative factors"

$=$ "Production reactions"/("Absorption reactions" + "Leakage")

The above idea is a basic approach to treat the behavior of neutrons in the neutron diffusion theory, which is called the "effective multiplication factor" with the notation "k_{eff}" for the adjustment factor k.

Here, in the neutron diffusion theory with one-energy group, the behavior of neutrons is expressed by using the effective multiplication factor k_{eff} as follows (Here, scattering reactions are neglected. For an explanation of neutron leakage, we refer the reader to other textbooks.):

$$-D\nabla^2\phi + \Sigma_a\phi = \frac{1}{k_{eff}}v\Sigma_f\phi, \tag{2.1}$$

where D is the diffusion coefficient, ϕ the neutron flux, Σ_a the absorption cross section, v the average number of neutrons triggered per fission reaction, and Σ_f the fission cross section. It is easy to understand that the left-hand side in Eq. (2.1) is the "decrease of number of neutrons," the right-hand side is the "increase of number of neutrons," and their adjusting factor is then $1/k_{eff}$.

Once again, interpreting the balance of neutrons in the system as the ratio of increase and decrease of number of neutrons, the balance of neutrons is a straightforward concept for understanding neutron multiplication. The k_{eff} in Eq. (2.1) is then written as follows:

$$k_{eff} = \frac{v\Sigma_f\phi}{-D\nabla^2\phi + \Sigma_a\phi}. \tag{2.2}$$

The change in the value of k_{eff} depends on the balance of denominator and numerator factors shown in Eq. (2.2) and can be interpreted as follows:

- When the number of neutrons generated by fission reactions is larger than those by absorption reactions and leakage, $k_{eff} > 1$.
- When the number of neutrons generated by fission reactions is equal to those by absorption reactions and leakage, $k_{eff} = 1$.
- When the number of neutrons generated by fission reactions is smaller than those by absorption reactions and leakage, $k_{eff} < 1$.

From the above interpretation, we think it is easy to judge that k_{eff} is closely related to the reactor condition. In other words, when the effective multiplication factor k_{eff} is unity, the reactor reaches a critical state. Moreover, when k_{eff} is smaller than unity, the reactor condition is called a subcritical state. If k_{eff} is larger than unity, the reactor condition is called a supercritical state.

(2) Generation of neutrons

The effective multiplication factor explained by the concept of the balance of neutrons was based on the assumption that the neutrons generated by fission reactions already existed in the reactor. Noteworthy is that the behavior of neutrons in this section is, however, different from that in the previous section ((1) Balance of neutrons), because the balance of neutrons is assumed to observe the system back to the origin of fission reactions and to recognize the existence of the neutron source that is an initial trigger of fission reactions.

In a system with the U-235 fuel and light-water moderators, fission reactions are impossible to be triggered with only the existence of fuels and moderators in the system. Here, seeds, as is called "source neutrons," are required for inducing an actual trigger of fission reactions.

Let us suppose that number of source neutrons S is emitted from a plutonium–beryllium (Pu-Be) neutron source. Also, let us focus on S emitted from the neutron source and observe how these neutrons move around in the core. Here, we introduce the ratio of change of neutrons (balance between increase and decrease in the number of neutrons) in the system introduced in the balance of neutrons.

In addition, let us denote the ratio of change as k_s (note that this is not the same as for the balance of neutrons, since a neutron source is existed in the core), number of source neutrons S is then inserted into the reactor, and neutrons cause repeatedly scattering, capture and fission reactions, and leak out of the reactor. When a series of events is considered an initial generation, one of two or three of the neutrons produced in the initial generation of fission reactions will contribute to fission reactions in the next generation. In short, before the initial trigger of fission reactions (zero generation), the number of source neutrons S is provided from the neutron source. After the first generation, the number of neutrons $k_s S$ will be produced.

Neutrons in a successive generation are multiplied by k_s again after the first generation, and the number of neutrons $k_s^2 S$ is produced after the second generation. Thus, a total number of neutrons are $S + k_s S + k_s^2 S + k_s^3 S + k_s^4 S + \cdots$ for the third, fourth generations, and so on up to infinity, including the initial number of source neutrons S. Then, if the total number of neutrons generated by fission reactions in all generations is F, that is, all generations are expanded to infinity, F can be expressed by using infinite geometric series as follows:

$$F = S + k_s S + k_s^2 S + k_s^3 S + k_s^4 S + \cdots = S\left(1 + k_s + k_s^2 + k_s^3 + k_s^4 + \cdots\right)$$
$$= \frac{S}{1 - k_s}. \tag{2.3}$$

Here, the multiplication of neutrons in the system is determined by the ratio of the number of source neutrons S and the total number of neutrons generated by fission reactions F. If the multiplication of neutrons is denoted as M, M can be expressed by using Eq. (2.3) as follows:

$$M = \frac{F}{S} = \frac{S}{1 - k_s} \cdot \frac{1}{S} = \frac{1}{1 - k_s}. \tag{2.4}$$

In Eq. (2.4), the multiplication M is diverging to infinity when k_s is approaching 1 (unity: a critical state); i.e., when k_s is less than 1 (a subcritical state), M converges to a certain value (saturation). In the approach to criticality experiment, to make an easy understanding of the multiplication M by increasing k_s, we transform Eq. (2.4) as follows:

$$\frac{1}{M} = 1 - k_s. \tag{2.5}$$

If Eq. (2.5) is defined as the inverse multiplication $1/M$, we observe that the inverse multiplication approaches zero as k_s approaches 1. Using the property, in the approach to criticality experiment, the parameter relating to k_s is set on the horizontal axis and the inverse multiplication on the vertical axis. The criticality of the core can be then determined by gradually bringing the reactor closer (larger) to the criticality (horizontal axis) and finding the point where the inverse multiplication (vertical axis) is zero (extrapolation).

2.1.2.2 Role of Neutron Source

The behavior of neutrons in the core was explained by introducing two concepts of "balance" and "generation" of neutrons, to understand easily the concept of neutron multiplication. The absence of neutron source was confirmed in the concept of the balance of neutrons, and conversely, in that of the generation of neutrons, the neutron source was present. Since a series of reactor physics experiments, except for control rod calibration experiment, could be conducted with the use of neutron source, an actual role of neutron source should be understood in the procedures of the experiments.

The neutron source used in UTR-KINKI is Pu-Be, and neutrons are generated by the $^9Be(\alpha, n)^{12}C$ reactions using particles emitted from Pu-239, which is one of the radioactive sources. The energy of the neutrons generated in the $^9Be(\alpha, n)^{12}C$ reactions is 4.5 MeV, and we can easily image that the neutrons are sufficiently fast because of their considerable energy. Considering that the half-life of Pu-239 is 24,110 years, Pu-Be can be used as a neutron source semipermanently.

Based on the neutron diffusion theory with one-energy group, how can we describe the behavior of neutrons in the core under the existence of neutron source? Neutrons have repeatedly scattering, capture and fission reactions, and leak out of the core. At the same time, source neutrons are continuously supplied from the neutron source. If the total number of neutrons supplied from the neutron source is S, the balance of neutrons can be then expressed as follows:

$$-D\nabla^2\phi + \Sigma_a\phi = \nu\Sigma_f\phi + S. \tag{2.6}$$

It can be observed that the ratio of change of neutrons in the presence of a neutron source is always constant; i.e., no significant change in the behavior of neutrons occurs in the reactor while neutrons are constantly supplied. As an aside, from Eq. (2.4), the right-hand side in Eq. (2.5) is positive at a subcritical state, $1 - k_s > 0$, resulting in $k_s < 1$. The multiplication of neutrons is also then, $M > 0$. The interpretation implies that the ratio of change of neutrons is less than 1 (unity) in the system where the neutron source is used; i.e., the steady-state reactor is always at a subcritical state under the existence of the neutron source in the core.

Here, we should always confirm whether a neutron source is in the system or not at the time of the experiment. This confirmation is a significant clue to determine whether the reactor is at a subcritical or not.

[Column] Extension of neutron diffusion equation to two-energy groups

The behavior of neutrons is conveniently described by the neutron diffusion theory with one-energy group as shown in Sect. 2.1.2.1, and here, let us consider what it looks like when the one-energy group is extended to the two-energy groups. Although Eq. (2.1) showed the neutron diffusion equation with one-energy group, please think about the reason why it is necessary to extend the equation from the one-energy group to the two-energy groups.

When we imagine the behavior of neutrons in a nuclear reactor, it is convenient to model the behavior in one-dimensional plate or two-dimensional cylinder to understand the physical phenomena qualitatively. Suppose the behavior is modeled in three-dimensional geometry. Then, we could model the behavior of neutrons in three-dimensional geometry more precisely than that in one-dimensional or two-dimensional geometry.

How the neutron energy is then? If the energy distribution (neutron spectrum) of neutrons in the reactor is dominated by fast neutrons, as in fast reactors, there is no serious obstacle to understanding the behavior of neutrons even if we assume the energy distribution of neutrons as one-energy group.

In UTR-KINKI, U-235 is used as a fuel, light water as a moderator and graphite as a reflector. When we imagine the behavior of neutrons in the reactor from this composition, the neutrons generated by U-235 fission reactions are almost fast neutrons, which are eventually changed into thermal neutrons through repeated reactions with light water as moderators, and graphite as reflectors playing a similar role as moderators in the reactor. Then, the energy of neutrons has a wide distribution of fast and thermal neutrons. If a wide range of energy is treated as one-energy group, the behavior of neutrons in a nuclear reactor cannot be accurately understood.

Therefore, we roughly divide the energy distribution of neutrons into two-energy groups, fast and thermal neutrons, and summarize the physical phenomena and interactions (reactions) of each group as well as the one-energy group diffusion equation. Here, the subscripted numbers represent the energy group ("1" is the fast group and "2" is the thermal group), Σ_a is the absorption cross section, respectively. Σ_s the scattering cross section, respectively. Σ_f the fission cross section, respectively. Let us consider the behavior of fast and thermal neutrons in terms of the effect of increasing or decreasing the number of neutrons in the core, as used in the concept of neutron multiplication, as follows:

(1) Fast neutron

(Effect of removal) Leakage outside the reactor, absorption reaction, downward scattering reaction from group 1 to group 2.

(Effect of production) Fission reaction (the only fast group is neutron source by fission).

The effects can be described by the diffusion equation using the neutron fluxes ϕ_1 and ϕ_2, the effective multiplication factor k_{eff} as follows:

$$-D_1 \nabla^2 \phi_1 + \left(\Sigma_{a,1} + \Sigma_{s,1\to 2}\right)\phi_1 = \frac{1}{k_{eff}}\left(\nu_1 \Sigma_{f,1}\phi_1 + \nu_2 \Sigma_{f,2}\phi_2\right), \qquad (2.7)$$

(2) Thermal neutron

(Effect of removal) Leakage out of the reactor and capture reaction. Here, upward scattering reactions from group 2 to group 1 are assumed to be absent.

(Effect of production) Neutron source by fission is not considered to be thermal neutron group.

The behavior of thermal neutrons, as well as that of fast neutrons, can be described by the diffusion equation as follows:

$$-D_2 \nabla^2 \phi_2 + \Sigma_{a,2}\phi_2 - \Sigma_{s\,1\to 2}\phi_1 = 0. \qquad (2.8)$$

Using Eqs. (2.7) and (2.8) as governing equations, for example, and giving the diffusion coefficient of each group and the cross section of each reaction, the neutron fluxes ϕ_1 and ϕ_2 of the fast and thermal groups, respectively, can be easily determined. The details of the two-energy-group diffusion equation are given in Refs. [1, 2].

2.1.3 Method of Measurement

Four possible ways are generally available to reach a critical state as follows:

(1) Increasing gradually the amount of fuel and reaching a critical state
(2) Increasing gradually the fuel concentration while keeping the core volume almost constant

(3) Loading the fuel as much as possible to reach a critical state in advance, and increasing gradually the water level to reach a critical state

(4) Loading the fuel as much as possible to reach a critical state in advance, and withdrawing gradually control rods to reach a critical state.

Among the four methods, the most common method is method (1), and, in this experiment, approach to criticality is performed by two methods: method (1) by increasing number of fuel plate; method (4) by operating control rods (to change the position). One of the two methods is, however, chosen depending on the experimental week, so you could observe the reactor carefully to confirm which method is used in the experiment.

2.1.3.1 Inverse Multiplication

The multiplication factor and its inverse value are obtained from neutron counts measured by the neutron detectors. Here, for the convenience of the experiment, we assume that the core is loaded with some fuel (hereinafter referred to as the initial core).

Let $k_{s,0}$ and $k_{s,i}$ the ratio of change of neutron multiplication in the initial core (subscript is "0.") and the core with the ith state of loading fuel (subscript is "i."), respectively. Since the neutrons generated by fission reactions correspond to the neutron fluxes $\phi_{s,0}$ and $\phi_{s,i}$ multiplied by the neutron source in the initial and i-th states, respectively, as shown in Eq. (2.3), we approximate them as follows:

$$\phi_{s,0} \approx \frac{S}{1 - k_{s,0}}, \tag{2.9}$$

$$\phi_{s,i} \approx \frac{S}{1 - k_{s,i}}. \tag{2.10}$$

When the state of the core varies from the initial core into i-th state core, the value of $\phi_{s,i}/\phi_{s,0}$ can be expressed by using Eqs. (2.9) and (2.10), as follows:

$$\frac{\phi_{s,i}}{\phi_{s,0}} \approx \frac{1 - k_{s,0}}{1 - k_{s,i}}. \tag{2.11}$$

Here, for any state i, the state of the initial core is the same; i.e., the numerator $1 - k_{s,0}$ of the right-hand side in Eq. (2.11) can be then regarded as being constant (this is an important assumption). Therefore, we can approximate Eq. (2.11) as follows:

$$\frac{\phi_{s,i}}{\phi_{s,0}} \approx \frac{1}{1 - k_{s,i}}. \tag{2.12}$$

The inverse value of Eq. (2.12) can be expressed as follows:

$$\frac{\phi_{s,0}}{\phi_{s,i}} \approx 1 - k_{s,i}. \tag{2.13}$$

Equation (2.13) has the same form as the inverse value of multiplication (inverse multiplication) $1 / M$ in Eq. (2.5), and the method for obtaining inverse multiplication $1 / M$ can be directly applied by attaining the inverse of the neutron flux ratio $\phi_{s,0}/\phi_{s,i}$ in Eq. (2.13).

Let us assume that the response (counting rate) per unit time obtained from the neutron detector is proportional to the neutron flux (this is a vital assumption). When obtaining the counts C_0 and C_i in states 0 and i, respectively, the ratio of the counting rates, C_0 / C_i, can be expressed from Eq. (2.13) as follows:

$$\frac{C_0}{C_i} \approx \frac{\phi_{k,0}}{\phi_{k,i}} \approx 1 - k_{s,i} \left(= \frac{1}{M_i} \right). \tag{2.14}$$

The obtained C_0/C_i is the inverse counting rate (inverse multiplication $1 / M_i$ in state i). In short, experimentally, the inverse counting rate of Eq. (2.14) corresponds to the inverse multiplication of Eq. (2.5) itself.

2.1.3.2 Settings of Detectors

To reach safely a critical state, detectors are set at several appropriate locations in the reactor. Here, the outline of approach to criticality by method (1) shown in Sect. 2.1.3 is described as follows:

As shown in Fig. 2.1, the decrease in the inverse of multiplication (vertical axis) against the increase in the amount of fuel to be loaded (horizontal axis) is plotted on a graph, and a straight line is obtained from the two most recent points and extrapolated to find the intersection point with the horizontal axis. The curve of this plot may be convex upward or convex downward, depending on the position of the detector. In the approach to criticality experiment, the detectors are arranged so that both curves can be obtained, and especially at a near-critical state, it is desirable to obtain a curve with a downward convex trend from the viewpoint of criticality safety. (see Fig. 2.1).

2.1.3.3 Fuel Loading

Fuel loading starts from the region of core center unless fuel is a liquid state. This is because the number of neutrons is highest near the center of the core, no matter what the core configuration is. If fuel is loaded here after the core is close to a critical state, the core will reach a critical state immediately.

The amount of the first fuel loaded is much less than the predicted critical mass. Let C_0 the total number of neutrons before the first fuel is loaded. After the first fuel is loaded, we wait for a while until the reactor power is constant (stable state), and measure the neutron counts in each condition (full insertion or full withdrawal of

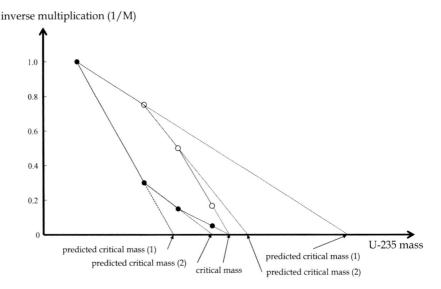

Fig. 2.1 Relationship between inverse multiplication $1/M$ and critical mass of U-235 fuel

control rods) at a stable state. Let C_1 the total number of neutrons obtained in this way for the first fuel loading, and the inverse counting rate for the first fuel loading (C_0/C_1) is obtained. Since the multiplication factor of neutrons is proportional to the counting value, the inverse counting rate C_0/C_1 is equivalent to the inverse multiplication factor k_0/k_1, where k_0 and k_1 are the multiplication factors of neutrons before and after the first fuel loading, respectively. As shown in Fig. 2.1, extrapolating the straight line obtained after repeated fuel loading (3 or 4 times), there is an intersection point between the straight line and the horizontal axis: the predicted critical mass. The additional fuel estimated from the predicted critical mass should not be loaded at once, and less than half of the predicted critical mass should be used as a guide for the next loading. Therefore, the amount of additional fuel should be reduced as the reactor is reaching a critical state, and the additional fuel loading should be handled with sufficient care.

2.1.3.4   Control Rod Operation

Approach to criticality experiment by operating control rods is, in principle, exactly the same as that by loading fuel. When all the control rods are withdrawn, the reactor is assumed to be at a supercritical state; i.e., there are enough fuel plates loaded in the core to make it critical. The control rod in UTR-KINKI is a plate absorber made of cadmium (Cd), so the control rod is sometimes referred to as a control plate or Cd plate in this text. (For details, please refer to Sect. 1.1.4.2).

The initial condition is when the control rod is at the lower limit, and the total value at that time is C_0. Here, the inverse counting rate of neutrons is the same as in the case of fuel loading, whereas the control rod position is on the horizontal axis, differing from the number of fuel plates (critical mass) shown in Sect. 2.1.3.3. The control rod position in the horizontal axis corresponds to the amount of fuel loaded in the case of fuel loading. The control rods move along the axial direction of the core, and the control rods are gradually withdrawn from the core, as in the case of fuel loading. For example, if C_1 is the value of neutron counts when the control rod is withdrawn by 10 cm from the lower limit (0 cm: center position in axial direction), the inverse counting rate is C_0/C_1, given by the initial inverse multiplication when the control rod is withdrawn. By repeating this process several times, the inverse multiplication curve resulted in the control rod operation is obtained. Also, "position of the control rod at a critical state" can be determined by extrapolating the straight line obtained from the last two points, since the horizontal axis is the position of the control rod. Although the method of the control rod operation is different from that of the fuel loading, the basic principle of measurement is the same in the two methods. In short, when the control rod position is gradually changed toward the upper limit, the predicted control rod position at a critical state can be estimated by measuring quantitatively the neutron multiplication.

2.1.4 Procedure of Measurement

The specific procedure for the approach to criticality experiment is described in this section. The schematic of the measurement system of the neutron detector used in the approach to criticality experiment is shown in Fig. 2.2, and data sheet in the approach to criticality experiment is shown in Fig. 2.3.

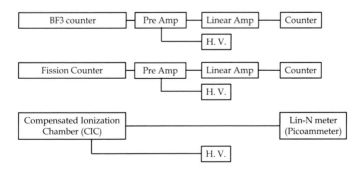

Fig. 2.2 Neutron measurement system used in approach to criticality experiment

No	Fuel Configuration	U-235 Mass (g)	Control Rod Position	Measured Current and Count Rate			Inverse Multiplication Factor		
				CIC (A)	FC (cps)	BF3 (cps)	CIC (A)	FC (cps)	BF3 (cps)
0	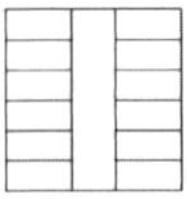		All inserted	$C_{0,1}$			$C_{0,1}/C_{0,1}$		
			All withdrawn	$C_{0,2}$			$C_{0,2}/C_{0,2}$		
1	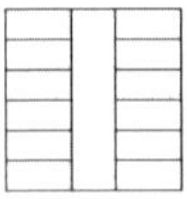		All inserted	$C_{1,1}$			$C_{0,1}/C_{1,1}$		
			All withdrawn	$C_{1,2}$			$C_{0,2}/C_{1,2}$		
2	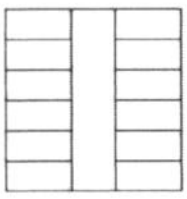		All inserted	$C_{2,1}$			$C_{0,1}/C_{2,1}$		
			All withdrawn	$C_{2,2}$			$C_{0,2}/C_{2,2}$		
3	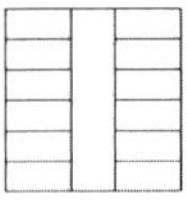		All inserted	$C_{3,1}$			$C_{0,1}/C_{3,1}$		
			All withdrawn	$C_{3,2}$			$C_{0,2}/C_{3,2}$		
4	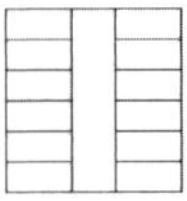		All inserted	$C_{4,1}$			$C_{0,1}/C_{4,1}$		
			All withdrawn	$C_{4,2}$			$C_{0,2}/C_{4,2}$		
5	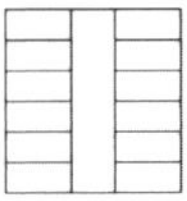		All inserted	$C_{5,1}$			$C_{0,1}/C_{5,1}$		
			All withdrawn	$C_{5,2}$			$C_{0,2}/C_{5,2}$		
6	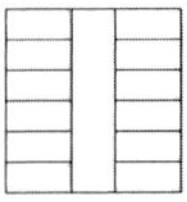		All inserted	$C_{6,1}$			$C_{0,1}/C_{6,1}$		
			All withdrawn	$C_{6,2}$			$C_{0,2}/C_{6,2}$		

Fig. 2.3 Data sheet in approach to criticality experiment

2.1.4.1   Approach to Criticality by Loading Fuel Plates

(1) Set the detector.
(2) Check the detector.

The detector has already been set, and its location should be checked. We will consider the relationship between the detector location and the concavity of the $1/M$ curve later.

The UTR-KINKI core that is a coupled core is divided into two parts (cores), but in principle, the core is possible to be considered as a single core.

(3) Check the conditions of the control rod and light-water moderators, and then, insert the neutron source.

A quantity of fuel (U-235) has already been loaded in the core, reaching a deep subcritical state.

In the state (denoted $i = 0$) before the first fuel is loaded, let C_0 the count under the existence of the neutron source.

(4) Measurements of the counting rate of neutrons are made as follows:
(a) When two control rods (Shim Safety Rod: SSR and Regulating Rod: RR) are inserted, withdrawing two Safety Rods (SRs #1 and #2) (denoted $j = 1$).
(b) When two control rods (SSR and RR) have been withdrawn, withdrawing two safety rods (SRs #1 and #2) (denoted $j = 2$).

In other words, two counting rates, $C_{0,1}$ and $C_{0,2}$, are obtained as the state before the fuel is loaded. (The measurements of $C_{0,1}$ and $C_{0,2}$ are also meant to check the effectiveness of the control rods at the same time in the process of the approach to criticality experiment.)

(5) All control rods are inserted when the measurement in (b) of step (4) is completed.
(6) The first fuel is loaded (fuel elements are carefully inserted one by one, paying attention to the increase of the counting rate (defined $i = 1$).
(7) The counting is conducted in the same way as in step (4). (Two neutron counting rates, $C_{1,1}$ and $C_{1,2}$, are obtained.)
(8) All control rods are inserted when the measurement in (b) of step (4) is completed.
(9) Take the fuel loading (mass of U-235) on the horizontal axis and the inverse counting rate (inverse multiplication) on the vertical axis of the graph, and plot the value of $C_{0,1}/C_{1,1}$ (and $C_{0,2}/C_{1,2}$).
(10) Assuming that the first plot point $C_{0,1}/C_{1,1}$ is normalized as unity ($= 1$), a straight line is obtained with the combined use of the next point obtained in step (9) and the first point, and the critical mass can be then predicted by extrapolating the line. Based on the results, the amount of fuel to be loaded next is determined. It is evident that the counting rate of neutrons increases as additional fuel is loaded, and let us observe that it takes time for the counting rate to converge (saturate).
(11) Repeat steps (6) to (10). ($i = 2, 3, 4$, etc., correspond to the number of fuel loadings.)
(12) At a near-critical state (e.g., $i = 4$), SRs #1 and #2 are withdrawn, and as a next step, following the withdrawal of SSR and RR. Here, the reactor is not yet at a critical state when the neutron source is withdrawn and the indicated value by the recorder shows a decreasing tendency. Conversely, when the indicated value reveals an increasing tendency with the fluctuation of reactor power, the neutron source is withdrawn. It is then observed whether the value is a constant value or not. The reactor reaches a critical state when the value of the recorder reveals a constant value, adjusting the positions of SSR and RR.

2.1.4.2 Approach to Criticality by Operating Control Rods

This section describes the procedure (step) of the approach to criticality experiment with the control rod operation.

(1) Set the detector.
(2) Check the detector.
(3) Insert the neutron source.

We know that fission chain reactions never occur at the positions where the SRs #1 and #2, the RR are at the lower limit, the SSR is at the upper limit. First, make sure that the positions of SRs #1 and #2 are at the upper limit, the RR is at the lower limit, and the SSR is at the lower limit (center of the Cd plate: 0%; 0 cm; denoted $i = 0$). The counting rate in the state $i = 0$ is C_0.

(4) The counting rate of neutrons is measured with SSR at the lower limit.
(5) Withdraw the SSR by 30% from the lower limit 0%. (Note the increase in the counting rate of neutrons; this is assumed to be $i = 1$).
(6) Take the position of SSR on the horizontal axis of the graph and the inverse counting rate (inverse multiplication) on the vertical axis, and plot the value of C_0/C_1.
(7) Assuming that the first plot point C_0/C_1 is normalized as unity ($= 1$), a straight line can be obtained with the combined use of the next point obtained in step (6) and the first point, and the SSR position that the reactor could reach a critical state is predicted by the linear extrapolation. Based on the result, the next step is available to determine a position where SSR should be withdrawn. It is evident that the counting rate of neutrons will increase as the SSR is withdrawn, and let us observe that it will take time for the counting rate to converge (saturate).
(8) Repeat steps (5) to (7) in the proportions of 50%, 70%, 80%, etc. ($i = 2, 3, 4$, etc., corresponding to the number of times where the control rod is withdrawn.)
(9) When the neutron source is withdrawn at a near-critical state (e.g., $i = 4$) and the counting rate shows a decreasing tendency of reactor power, the reactor is not yet at a critical state. Conversely, the value indicated by the recorder shows an increasing tendency with the fluctuation of reactor power, when the neutron source is withdrawn from the core at this time. Then, we observe whether the indicated value is a constant value or not. When the indicated value of the recorder shows a constant value, the reactor is at a critical state.

2.1.5 Discussion

(1) Does the inverse multiplication curve show an upward convexity, a downward convexity, or a shape close to a straight line? In particular, let us observe the position of the detector and the positional relationship with the neutron source, and consider why different shapes are obtained depending on the position of the

detector, including physical reasons (Only the case of "Approach to Criticality by Loading Fuel Plates" in Sect. 2.1.4.1 will be considered.).

(2) Let us determine whether the critical mass obtained from each curve can be regarded as appropriate values.

(3) Assuming that there is only one detector to be used in the experiment, let us examine from the viewpoint of criticality safety which detector should be chosen to predict the critical mass (positions of control rods at a critical state) of the core appropriately, i.e., which of the above inverse multiplication curves is suitable for predicting the critical mass.

2.2 Control Rod Calibration Experiment

2.2.1 Purpose

The reactivity of a reactor depends on various factors, including the amount of fuel, the temperature of moderators, the positions of control rods, etc. In a low-power reactor like UTR-KINKI, the reactivity that the reactor has when all the control rods are fully withdrawn is called the "excess reactivity." When all control and safety rods are withdrawn fully from a criticality state (for $k_{eff} = 1$, reactivity in Eq. (1.9) is $\rho = 0$), the reactor is reaching at a supercritical state and the reactor power increases. If the reactivity of the reactor at this time is defined as ρ_{excess}, the change in the position of the control rod causes the reactivity change of the reactor, $0 \rightarrow \rho_{excess}$. Therefore, the change in the position of the control rod is said to have "an equivalent reactivity value of ρ_{excess}," which is an index of the effect of the control rod on the reactor. In the control rod calibration experiment, the "equivalent reactivity" of each control rod is obtained in this way.

The control rod calibration experiment is carried out just after the reactor reaches a critical state for the first time, to confirm the essential characteristics of control rods that play a significant role in the safety of the reactor. The reactivity worth of the control rod obtained by the control rod calibration experiment is used as a reference for various reactivity measurements, such as the reactivity of irradiated materials inserted in the core: sample reactivity worth. In the control rod calibration experiment, excess reactivity and control rod reactivity worth in the UTR-KINKI core can be determined using the positive period method and the rod drop method, respectively, which are typical reactivity measurement methods. The experiment aims to deepen the understanding of the role of control rods and nuclear safety of nuclear reactors by comparing and discussing the measured reactivity values.

2.2.2   Structure of Control Rod

The control rods used in UTR-KINKI (strictly speaking, they are control plates made of Cd) are described in terms of their positions in the core (see Sect. 1.1.4). Figure 2.4 shows a longitudinal cross section of the core in which the control rod is inserted. Looking at Fig. 2.4 carefully, you will see that the control rods in UTR-KINKI are not located next to most of the fuel region, as is usually the case in research reactors.

As shown in Fig. 2.4, when the control rod is fully inserted into the core (left-side control rod), the position of the center of the absorber (control plate) corresponds to that of the center of the fuel region (fuel plate). The control rod is then at the lower limit. On the other hand, when the control rod is fully withdrawn (right-side control rod), the center of the absorber is about 10 cm higher than the upper height of the fuel plate, and here, the control rod is at the position of the upper limit. The axial relationship between control rods and fuel is shown in Fig. 2.4.

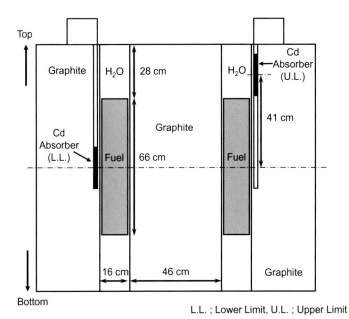

Fig. 2.4 Axial view of positional relationship between control rod (safety rod) and fuel plate in UTR-KINKI

2.2.3 *Method of Measurement*

2.2.3.1 Reactivity Measurement by Positive Period Method

Of several methods of measuring the reactivity, the most standard "positive period method" is used to measure the relatively small positive reactivity.

As shown in Fig. 2.5, the reactor power increases exponentially when a positive reactivity is added to the reactor at a critical state. State a is an example of the case where the reactivity ρ_a is large, and state b is that of the case where the reactivity ρ_b is small ($\rho_a > \rho_b$). The doubling time T_{2a} for state a is the same at any time, and the doubling time T_{2b} for state b is also the same. Here, the doubling time is the time it takes for a reactor power to double. As shown in Fig. 2.5, the larger the reactivity, the larger the rate of increase of the reactor power and the shorter the doubling time: $T_{2a} < T_{2b}$.

The increase of the reactor power p can be then expressed as follows:

$$p(t) = p_0 e^{\frac{t}{T}}, \tag{2.15}$$

where p_0 is the reactor power at the time when the measurement is started ($t = 0$), t the time from the start of measurement and T the reactor period. Here, the reactor power is notably proportional to the number of neutrons generated by fission reactions.

The relationship between the reactor period T (Sect. 1.2.8.3) and the reactivity ρ (Sect. 1.2.7) is determined by the inhour equation as follows (The details are omitted. Please refer to Ref. [3]):

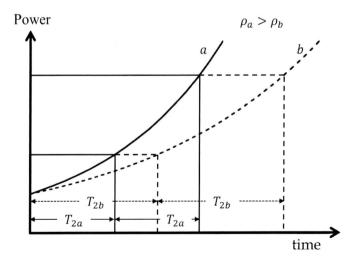

Fig. 2.5 Relationship between reactivity and reactor power

$$\rho = \frac{\ell_0}{T + \ell_0} + \frac{T}{T + \ell_0} \sum_{i=1}^{6} \frac{\beta_{eff,i}}{1 + \lambda_i T}, \tag{2.16}$$

ℓ_0 Prompt neutron lifetime of prompt neutrons (1.605×10^{-4} s by MVP3.0 [4] with JENDL-4.0 [5] for UTR-KINKI),
$\beta_{eff,i}$ Effective delayed neutron fraction of the i-th delayed neutron precursor,
λ_i Decay constant of the i-th delayed neutron precursor,

where the delayed neutron parameters of UTR-KINKI are given in Table 1.4.

The reactivity ρ can be obtained by substituting the reactor period T into Eq. (2.16), together with the parameters in Table 1.4.

Since it is difficult to directly measure the reactor period T (time for the reactor power to increase by e times) in actual measurement, when the doubling time T_d is obtained at the experiment, the reactor period T is determined by the following equation:

$$T = \frac{T_d}{\ln 2} = \frac{T_d}{0.693}. \tag{2.17}$$

The procedure for measuring the reactivity by the positive period method is as follows, referring to the data sheet shown in Fig. 2.6:

(1)   The reactor is kept at a critical state with low power.
(2)   The control rod is withdrawn, and a relatively small positive reactivity is added.
(3)   Since the reactor power increases exponentially, the doubling time T_d at which the reactor power doubles is measured by using a stopwatch.
(4)   Using Eq. (2.17), calculate the reactor period T from the doubling time T_d. Substitute the reactor period T and the delayed neutron parameters in Table 1.4 into the inhour equation in Eq. (2.16) to obtain the reactivity ρ.

[Column] Delayed criticality and prompt criticality

Let us consider the magnitude of the reactor period using the following discussion: Suppose that the reactor condition varies from a critical state $\left(k_{eff} = 1.000\right)$ to a supercritical state $\left(k_{eff} = 1.001\right)$ while UTR-KINKI is in operation. Assuming that the prompt neutron lifetime is approximately 0.0001 s (10^{-4}), under no existence of delayed neutrons, as shown in Sect. 2.2.3, $\ell_0 = 1.605 \times 10^{-4}$s, the reactor period T is about $T = \ell_0/(k_{eff} - 1) = 10^{-4}/10^{-3} = 0.1$ s from Eq. (1.14). Using Eq. (2.15), the number of neutrons increases by $e^{1/0.1} = 22,000$ times in one second. Then, it is evident that the neutron multiplication is so fast that the reactor power cannot be controlled very well.

Conversely, the situation is very different when delayed neutrons are present. The evaluation equation is slightly different (see Ref. [3] for details). If we substitute $k_{eff} = 1.001$ and the delayed neutron parameters in Table 1.4 into Eq. (1.14), we obtain $T = \left\{k_{eff}/(k_{eff} - 1)\right\} \cdot (\beta/\lambda), \{1.001/(1.001-1.000)\} \cdot (0.007342/0.0746)$

Control Rod Calibration: Data Sheet

Date: _____   Time: _____

Central Stringer: _____ cm inserted   Moderator Temperature: _____ °C

1. Reactivity Measurement by the Positive Period Method

1) Regulating Rod (RR)

Exp. 1

			$\times 10^{-10}$ A	$\times 10^{-9}$ A	$\times 10^{-8}$ A
_____ %	2.5→5.0				
SSR: _____ % ↓	3.0→6.0				
_____ %	4.0→8.0				

Exp. 2

			$\times 10^{-10}$ A	$\times 10^{-9}$ A	$\times 10^{-8}$ A
_____ %	2.5→5.0				
SSR: _____ % ↓	3.0→6.0				
_____ %	4.0→8.0				

2) Shim Safety Rod (SSR)

			$\times 10^{-10}$ A	$\times 10^{-9}$ A	$\times 10^{-8}$ A
_____ %	2.5→5.0				
RR: _____ % ↓	3.0→6.0				
_____ %	4.0→8.0				

2. Reactivity Measurement by the Rod Drop Method

1) Shim Safety Rod (SSR) _____ % → 0%

   (RR = 0%)   Count Rate at Critical State: _____ cps

   Integrated Counts after Rod Drop: _____ counts

2) Safety Rod (SR)   (#1 , #2)   100% → 0%

   (RR = 0%)   Count Rate at Critical State: _____ cps

   Integrated Counts after Rod Drop: _____ counts

Fig. 2.6   Data sheet in control rod calibration experiment

= 98.52 s. The reactor power after one second is only $e^{1/98.52} = 1.0102$ times, and, as a result, the reactor power is possible to be controlled with a sufficient margin.

In this way, the state in which a reactor is at a critical state with both prompt and delayed neutrons is called delayed critical, whereas the state in which a reactor is critical with only prompt neutrons is called prompt critical.

2.2.3.2 Reactivity Measurement by Rod Drop Method

One of the methods for obtaining a relatively large negative reactivity is the control rod drop method (rod drop method). In the rod drop method, when a control rod is inserted into the reactor at once, the reactivity of the control rod is obtained soon. As shown in Fig. 2.7, when a control rod is inserted into the reactor at the time t_0, the degree of decrease in the counting rate of neutrons after the insertion depends on the reactivity of the inserted control rod, even though the counting rate (power) of the neutron detector before the insertion is the same as that after the insertion. When the reactivity of the control rod to be inserted is large, the power "a" decreases rapidly, as shown in Fig. 2.7.

The ratio $\frac{n_0}{\int_0^\infty n(t)dt}$ of the pre-insertion counting rate n_0 and the post-insertion total counting rate of the control rod $\int_0^\infty n(t)dt$ is proportional to a negative reactivity $(-\rho)$ as follows:

$$-\rho = \frac{n_0}{\int_0^\infty n(t)dt} \sum_{i=1}^{6} \frac{\beta_{eff,i}}{\lambda_i}, \qquad (2.18)$$

where $n(t)$ is the counting rate of neutrons at the time t after the start time of measurement t_0 when the control rod is dropped.

Equation (2.18) was given by Hogan [6] (integral method) and has been often used to determine the integral reactivity of control rods. The reactivity value of a dropped control rod can be obtained by substituting the delayed neutron parameters in Table 1.4, the measured results of n_0 and $\int_0^\infty n(t)dt$ into Eq. (2.18).

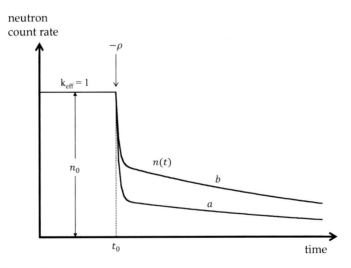

Fig. 2.7 Relationship between reactivity and output change by rod drop method

The procedure for measuring the reactivity by the rod drop method is as follows, referring to the data sheet in Fig. 2.6:

(1) The reactor is kept at a critical state and a reactor power (n_0) of about 0.1 W (steady state).
(2) The counting rate of neutrons n_0 (s^{-1}; cps: counts per second) is actually obtained from the measurement values of m (cps) that is corrected for the dead time of the measuring instrument (see [Column] "Dead time correction").
(3) The control rod is dropped rapidly to add a relatively large negative reactivity in the core.
(4) The value of the time integral of the neutron counts $\int_0^\infty n(t)\,dt$ is measured after the control rod is dropped.
(5) The negative reactivity $(-\rho)$ is obtained by substituting the measured n_0 and $\int_0^\infty n(t)\,dt$, and the delayed neutron parameters in Table 1.4 into Eq. (2.18).

2.2.4 Discussion

(1) Let's check whether the excess reactivity obtained by the positive period method was an appropriate value within the critical limits (or reactor limits) of UTR-KINKI (Table 1.1 in Chap. 1).
(2) Do the results of reactivity of the control rods obtained by the rod drop method satisfy the limiting conditions of reactivity (critical limits or reactor limits) of UTR-KINKI, as shown in Table 1.1? If the reactivity of each of several control rods obtained by the rod drop method seems to be different from each other, let us consider the reason why the results are different even though the same method was used.

[Column] Reactivity calibration curve of control rods

The main role of control rods is to control fission chain reactions for keeping the reactor at a critical state and adjusting the reactor power. Here, if a control rod is inserted into the reactor, neutrons are absorbed, while fission chain reactions are suppressed. Conversely, fission chain reactions are activated when the control rod is withdrawn from the reactor. Before the operation of the reactor, it is crucial to know in advance how much (positive or negative) reactivity is added to the reactor by the control rod operation.

In the control rod calibration experiment in UTR-KINKI, when the control rods are moved for the whole stroke from the upper limit (full withdrawal) to the lower limit (full insertion) (or from the lower limit to the upper limit), the reactivity of the control rod can be obtained by using the positive period method and/or the rod drop method. The control rod calibration experiment is then conducted to know the reactivity to shut down the reactor safely.

By using the rod drop method, the response of the control rod to a full stroke of the control rod was obtained. Meanwhile, what about the response to the reactor power when the control rod is partially moved from a certain position to another? Is the amount of reactor power change always constant at any position for the stroke of control rod movement? Surprisingly, the answer is that the reactor power change is not constant. As shown from the reactor structure described in Chap. 1 (cross-sectional view of UTR-KINKI in Fig. 1.4), the control rod moves in the region adjacent to the fuel. From the positional relationship between the control rod and the fuel (Fig. 2.4), you can easily imagine that the effectiveness of the control rod depends strongly on neutron density of the fuel region along an axial direction. Without investigating the effect of control rods on the direction of movement, the reactor operation cannot be conducted safely. Therefore, the control rod calibration experiment is considered very important in reactor physics experiments, in accordance with the approach to criticality experiment.

[Column] Dead time correction

When radiation is incident on a measuring instrument, there is a short time when the instrument cannot measure the next incident radiation due to the recovery of the instrument and data processing. This time is called the "dead time." If the measured counting rate is m (s^{-1}: cps) and the dead time is τ (s), the correct measurement time of the instrument in one second is $(1 - m\tau)$. Therefore, the actual counting rate N_0 (cps), which takes into account the counting off due to the dead time in the measured value, is expressed by the following equation:

$$N_0 = \frac{m}{1 - m\tau}. \tag{2.19}$$

Typically, the dead time of the fission counter used for the rod drop method measurement is about 1.4 μs.

2.3 Subcriticality Measurement Experiment

The subcriticality is the index that quantifies "how much margin is left in the system from the critical state $k_{eff} = 1$?" and is the quantity defined by the negative reactivity $(-\rho) = (1 - k_{eff})/k_{eff}$.

The criticality state of a nuclear reactor is defined as a state in which neutron production by fission reactions is balanced by neutron annihilation due to absorption reactions in the system or leakage outside the system. When the reactor core is maintained at a steady state for a long period without an external neutron source (Pu-Be start-up neutron source) inserted into the core, the core is judged to be at a critical state.

Here, if the reactor core is at a subcritical state $(k_{eff} < 1)$, the number of neutrons in the target system will be zero under no existence of external neutron source in the core. In this case, if the counting rate of neutrons measured by a neutron detector is zero, the core may be judged to be at "a subcritical state." It is, however, impossible to determine the magnitude of subcriticality from the information that the counting rate of neutrons is zero. Therefore, to investigate the magnitude of subcriticality $(-\rho)$ in a subcritical core, the number of neutrons is measured with the help of the external neutron source that induces the trigger of fission chain reactions, and the magnitude of subcriticality $(-\rho)$ is then analyzed.

Although several methods have been proposed to measure the subcriticality [6–10], in this section, we focus on measurement methods that can be used in UTR-KINKI and introduce several subcriticality measurement methods.

2.3.1 Neutron Source Multiplication Method

2.3.1.1 Principle of Measurement

Consider the case where the reactor is at a subcritical state and the reactor power is constant (no change with time) due to an external neutron source. In this case, if the number of neutrons and the number of delayed neutron precursors are n_0 and C_0, respectively, the point-reactor kinetics equations with the addition of the neutron source are shown in Eqs. (1.10) and (1.11), considering that the reactor is at a steady state. Then, Eqs. (1.10) and (1.11) can be expressed as follows, respectively, neglecting the change in time t:

$$\frac{dn_0}{dt} = \frac{\rho_0 - \beta}{\Lambda} n_0 + \lambda C_0 + S = 0, \tag{2.20}$$

$$\frac{dC_0}{dt} = \frac{\beta}{\Lambda} n_0 - \lambda C_0 = 0, \tag{2.21}$$

where ρ_0 is a negative value of reactivity and S neutron source intensity.

From Eq. (2.21), the following equation can be obtained:

$$\lambda C_0 = \frac{\beta}{\Lambda} n_0. \tag{2.22}$$

Here, Eq. (2.22) can be rewritten using $\Lambda = \ell/k_{eff}$ and $\rho_0 = 1 - \frac{1}{k_{eff}}$ in Eq. (1.9) as follows:

$$\lambda C_0 = \frac{\beta k_{eff}}{\ell} n_0 = \frac{\beta}{\ell(1 - \rho_0)} n_0. \tag{2.23}$$

Inserting Eq. (2.23) into Eq. (2.20), we obtain the following equation for n_0, and noteworthy is that ρ_0 has a negative value:

$$n_0 = -\ell(1 - \rho_0)\frac{S}{\rho_0} = \ell\left(1 - \frac{1}{\rho_0}\right)S. \tag{2.24}$$

Assuming that the reactor is at a near-critical state, and using $1 - \rho_0 \approx 1$ and $\Lambda \approx \ell$, the n_0 in Eq. (2.24) can be approximated as follows:

$$n_0 \approx \Lambda \frac{S}{(-\rho_0)}. \tag{2.25}$$

Equation (2.25) shows that the power of a subcritical reactor is proportional to the neutron source intensity S and inversely proportional to $(-\rho_0)$. The method for measuring the reactivity in the subcritical state above is called the "neutron source multiplication method."

How can we measure the subcriticality $(-\rho_0)$ in an actual reactor using this principle? We will discuss this question by modifying Eq. (2.24).

The reactor is at a known subcritical state where the multiplication factor is k_0, and the detection efficiency of the detector set in the core is ε. Equation (2.24) is transformed into the following expression for the multiplication factor k_0:

$$n_0 = \frac{\ell \varepsilon S}{1 - k_0}. \tag{2.26}$$

If the multiplication factor k_0 and the counting rate n_0 measured by the detector are the known values, the inverse value of the unknown $\ell \varepsilon S$ in Eq. (2.26) can be obtained as follows:

$$\frac{1}{\ell \varepsilon S} = \frac{1}{n_0(1 - k_0)}. \tag{2.27}$$

Next, when the reactor condition varies from the known subcritical state (where the multiplication factor is k_0) to the unknown subcritical state (where the multiplication factor is k_1), the multiplication factor k_1 is expressed as the same formation as Eq. (2.26). If the counting rate from the detector at this time is n_1, the multiplication factor k_1 can be obtained by using Eq. (2.27) as follows:

$$k_1 = 1 - \frac{n_0}{n_1}(1 - k_0). \tag{2.28}$$

From Eq. (2.28), the unknown multiplication factor k_1 obtained by measuring the counting rate n_1.

Then, using Eq. (1.9), the known subcriticality $(-\rho_0)$ can be expressed as $(-\rho_0) = \frac{1-k_0}{k_0}$, and the unknown multiplication factor k_1 in Eq. (2.28) can be expressed by using the known subcriticality $(-\rho_0)$ as follows:

$$k_1 = 1 - \frac{n_0}{n_1}\left(\frac{-\rho_0}{1 - \rho_0}\right). \tag{2.29}$$

Since the unknown subcriticality $(-\rho_1)$ can be expressed by using Eq. (1.9) as $(-\rho_1) = \frac{1-k_1}{k_1}$, and the multiplication factor k_1 in Eq. (2.29), $(-\rho_1)$ can be finally expressed as follows:

$$(-\rho_1) = \frac{1 - k_1}{k_1} = \frac{1}{k_1} - 1 = \frac{1}{\frac{n_1}{n_0}\left(\frac{1-\rho_0}{-\rho_0}\right) - 1}, \tag{2.30}$$

where the unit of $(-\rho_0)$ is needed to be converted into $\Delta k/k$, when obtaining $(-\rho_0)$ in the units of pcm $(10^{-5}\Delta k/k)$ or $\%\Delta k/k$ $(10^{-2}\Delta k/k)$. Then, the unit of $(-\rho_1)$ indicates $\Delta k/k$.

2.3.1.2   Method of Measurement

Suppose that the excess reactivity ρ_{excess} and the reactivity worth of the control rod $(-\rho_{rod})$ obtained by the positive period method and the rod drop method, respectively, two values of ρ_{excess} and $(-\rho_{rod})$ are given in advance by the control rod calibration experiment. Let the subcritical state obtained by the drop method known (subscript is "0"), and consider the unknown state when another control rod is dropped from this state (subscript is "1").

The subcriticality of the unknown state is obtained by the neutron source multiplication method using the following procedures:

(1)   The ρ_0 was obtained in advance, using the excess reactivity ρ_{excess} and the reactivity of the control rod $(-\rho_{rod})$ ($\rho_0 = \rho_{excess} - (-\rho_{rod})$).

(2)   Setting the neutron source in the reactor, we wait for a while until the reactor power is constant (stable) in a known subcritical state with one control rod inserted.

(3)   Using a detector in the reactor, such as a fission chamber (FC) in the startup system, the number of neutrons is counted for 100 s and repeated, for example, five times. The average value n_0 of the counting rate is obtained.

(4)   Keep the neutron source and insert another control rod at the known subcritical state. When the reactor is at an unknown subcritical state, we wait for a while until the reactor power is at a stable state.

(5)   Using the same method for the known subcritical state, obtain the average value n_1 of the counting rate in the unknown subcritical state.

(6)   The unknown subcriticality $(-\rho_1)$ can be attained by substituting the counting rates n_0 and n_1, and the known subcriticality $(-\rho_0)$.

2.3.1.3   Discussion

Let's discuss the results of the reactivity (subcriticality) in an unknown subcritical state $(-\rho_1)$ according to the following point:

The reactivity of another control rod inserted into the known subcritical state is assumed to be attained in advance. Next, the subcriticality of the unknown state can be easily estimated from the sum of the subcriticality of the known state and the reactivity of another inserted control rod. Finally, we shall use this summation as a reference value for the experiment.

(1) How about the measurement accuracy (relative difference between the reference and measured values, or relative error) of the $(-\rho_1)$ by the neutron source multiplication method when comparing with the reference value? If a significant difference is found, what is the cause of the difference?

(2) Let us consider whether there is any difference between the neutron source multiplication method and the other methods introduced in Sect. 2.3, including the source jerk method and the inverse kinetics method.

(3) Let us judge whether the measured results are subcritical enough to be regarded as a near-critical state, which is an assumption of using the neutron source multiplication method.

(4) Are there any significant differences in the measurement results when multiple detectors are used? In short, let's discuss the difference between the measurement results at several detector positions. In addition to the positions of the detectors, we should also discuss spatial relationships between the detector and the neutron source; the detector and the fuel region; the structural materials around the detector.

2.3.2 Source Jerk Method

2.3.2.1 Principle of Measurement

Let us assume that the external neutron source is withdrawn instantaneously in a steady state of a subcritical core with an external neutron source. By measuring neutron counts until delayed neutron precursors in the subcritical core are completely decayed to be zero, the subcriticality $(-\rho)$ can be experimentally analyzed. The above-mentioned subcriticality measurement technique is called the "source jerk method" [6].

First, let us suppose that an external neutron source with the intensity S is inserted into a subcritical core with the subcriticality $(-\rho)$. Based on a simplified point kinetics equation with one neutron energy group and one delayed neutron precursor group, the numbers of neutrons and delayed neutron precursors, $n(t)$ and $C(t)$, can be, respectively, described as follows:

$$\frac{dn(t)}{dt} = \frac{\rho - \beta}{\Lambda} n(t) + \lambda C(t) + S, \tag{2.31}$$

$$\frac{dC(t)}{dt} = \frac{\beta}{\Lambda} n(t) - \lambda C(t). \tag{2.32}$$

Compared with Eq. (1.10), the major modification is that the external source term S is added to the right-hand side in the case of Eq. (2.31) because the S neutrons per second are emitted due to the decay of the external source. When a sufficient time is passed after the insertion of the external neutron source, the numbers of neutrons and delayed neutron precursors become constant values, n_0 and C_0, respectively. The analytical solutions of n_0 and C_0 in the steady state can be obtained by considering the time derivatives in Eqs. (2.31) and (2.32) are zero, respectively:

$$n_0 = \Lambda \frac{S}{(-\rho)}, \tag{2.33}$$

$$C_0 = \frac{\beta}{\lambda \Lambda} n_0 = \frac{\beta}{\lambda} \frac{S}{(-\rho)}. \tag{2.34}$$

Namely, these constant values of n_0 and C_0 before withdrawing the external neutron source are proportional to the magnitude of the neutron source strength S and inversely proportional to the subcriticality $(-\rho)$.

Next, let us suppose that the external neutron source is instantaneously withdrawn from the subcritical core at time $t = 0$; i.e., the source intensity S can be rapidly reduced to zero. In this case, as shown in Fig. 2.8, the prompt neutron component of the fission chain reaction due to the external neutron source disappears instantaneously. Note, however, that there are still delayed neutron precursors which are accumulated in the subcritical system before withdrawing the external neutron source. Therefore, the neutron count rate after withdrawing the external neutron source does not instantaneously become to be zero, although the count rate promptly decreases to a certain count rate n_d due to the decay of delayed neutron precursors. Such a physical phenomenon of the neutron count rate in a neutron multiplication system after an instantaneous change (e.g., withdrawal of external neutron source) is called "prompt jump." After the prompt jump, the number of neutrons in the target subcritical system decreases exponentially with time until the delayed neutron precursors in the core completely disappear due to their decays.

Fig. 2.8 Time variation of neutron count rate in the source jerk method

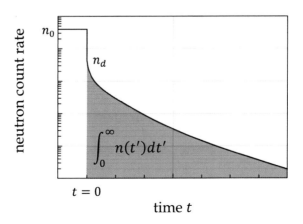

Here, the numbers of neutrons and delayed neutron precursors, $n(t)$ and $C(t)$, after withdrawing the external neutron source decrease according to the point kinetics equation with $S = 0$, i.e., Eqs. (1.10) and (1.11). Furthermore, $n(t)$ and $C(t)$ become to be zero after a sufficiently long time passes, i.e., $\lim_{t \to \infty} n(t) = \lim_{t \to \infty} C(t) = 0$. To obtain the subcriticality estimation formula in the source jerk method, both sides of Eqs. (1.10) and (1.11) are integrated within the range of $0 \le t \le \infty$. Then, both time-integrated equations are summed up and further transformed by substituting Eq. (2.34) into the term C_0. Finally, the following formula is obtained to estimate the subcriticality $(-\rho)$ from the measured neutron count rates:

$$-\rho = \frac{n_0}{\int_0^\infty n(t')dt'} \frac{\beta}{\lambda}. \tag{2.35}$$

If delayed neutrons with six groups are considered, the more rigorous formula can be obtained to estimate $(-\rho)$ as follows:

$$-\rho = \frac{n_0}{\int_0^\infty n(t')dt'} \sum_{i=1}^{6} \frac{\beta_{\text{eff},i}}{\lambda_i}. \tag{2.36}$$

Equation (2.36) is just the same formula as the integral method in the rod drop experiment [6].

[Column] Inherent neutron sources in nuclear fuel

Even if the external neutron source is completely withdrawn in the source jerk method, the neutron count rate may not be just zero due to another contribution of the "inherent neutron" in the nuclear fuel. For example, in the case of UTR-KINKI, the uranium–aluminum (U-Al) alloy is used as a nuclear fuel material. Because U-235 is a radioactive nuclide, secondary neutrons are emitted by the (α, n) reaction which rarely occurs when the Al-27 nuclide collides with an α particle emitted by the α-decay of uranium. As another example, in the case of low-enriched uranium, neutrons are rarely emitted by the spontaneous fission reaction of U-238. These reactions in the nuclear fuel can be regarded as very weak neutron sources that cause fission chain reactions [7]. Therefore, even if a long time passed after completely withdrawing the external neutron source, an extremely low neutron count rate n_∞ is maintained by the inherent neutron source. If n_∞ is nearly equal to zero, the impact of the inherent neutron source on the source jerk method is negligible. Otherwise, the subcriticality $(-\rho)$ can be more accurately estimated by considering the background neutron count rate n_∞ due to the inherent neutron source, i.e., by correcting the contribution of n_∞ as follows [8]:

$$-\rho = \frac{n_0 - n_\infty}{\int_0^\infty (n(t') - n_\infty)dt'} \sum_{i=1}^{6} \frac{\beta_{\text{eff},i}}{\lambda_i}. \tag{2.37}$$

2.3.2.2 Method of Measurement

To measure the subcriticality using the source jerk method for a target subcritical core, the experimental procedures are described as follows:

(1) A neutron detector (e.g., BF-3 detector) is installed in a hole of the graphite reflector to measure the neutron count rate using a typical nuclear instrumentation system as shown in Fig. 2.2.

(2) A target subcritical core is configured by an arbitrary pattern of inserting control rods (two safety rods: SRs, a shim safety rod: SSR, and a regulating rod: RR).

(3) A startup Pu-Be neutron source is inserted into the target subcritical core. Note that the effective source intensity (or the inserted axial position of the Pu-Be source) should be appropriately adjusted to reduce the dead time effect on the neutron count rate n_0 in the steady state after inserting the Pu-Be source.

(4) The target core is maintained without any change until the neutron count rate becomes a constant value.

(5) For example, the measurements of the neutron count during 100 s are repeated five times. From the five measured counts, the sample mean of neutron count rate n_0 (count per second, cps) before withdrawing the Pu-Be source is calculated.

(6) The neutron source is withdrawn instantaneously, and the time integral of the neutron count rate after the withdrawal, $\int_0^T n(t')dt'$, is cumulatively measured. Here, the total measurement time T for the time integral depends on the magnitude of the subcriticality $(-\rho)$. A typical value of T is approximately 300 to 500 s.

(7) If possible, the background neutron count rate n_∞ after withdrawing the Pu-Be source is measured in a similar way as step (5) to check whether the approximation of $n_\infty \approx 0$ is applicable.

(8) The target subcriticality $(-\rho)$ is estimated by substituting measured n_0 and $\int_0^\infty n(t')dt'$ into Eq. (2.36).

2.3.2.3 Discussion

Let us discuss the experimental results of subcriticality $(-\rho)$ using the source jerk method from the following viewpoints:

(1) As can be seen from the experimental result, the time variation of the count rate after withdrawing the external neutron source decreases exponentially according to the decay of the delayed neutron precursors. Based on the half-life of Br-87 (i.e., 55.6 s shown in Table 1.2), let us discuss whether the total measurement time $T = \sim 300$ s is sufficiently long.

(2) Let us consider the following two subcritical cores: (a) a shallow subcritical (or near-critical) core where the subcriticality $(-\rho)$ is close to zero, and (b) a deep subcritical case where $(-\rho)$ is large. Based on Eq. (2.36), which of them is

more accumulated delayed neutron precursors before withdrawing the external neutron source?

(3) Before the measurement using the source jerk method, let us measure the excess reactivity ρ_{excess} and the control rod worth $\rho_{rod,i} < 0$ using the positive period method and the control rod drop method, respectively, in advance. Then, based on the control rod pattern for the target subcritical core, let us evaluate the reference value of the subcriticality $(-\rho_{ref})$. By comparing with the reference value $(-\rho_{ref})$, is there a significant difference in $(-\rho)$ between the source jerk method and the reference? If any, what is the major reason?

(4) In the case of the source jerk method, the external neutron source is instantaneously withdrawn. Instead of withdrawing the external neutron source, let us discuss whether the subcriticality can be estimated by instantaneously dropping (or inserting) the external neutron source. Based on the point kinetics equation, derive the theoretical formula to estimate the subcriticality using the "source drop method."

2.3.3   Inverse Kinetics Method

2.3.3.1   Principle of Measurement

Let us change the reactivity $\rho(t)$ by inserting or withdrawing the control rod position in a critical or subcritical core. According to the positive or negative reactivity change, the number of neutrons, $n(t)$, increases or decreases with time. In other words, the total number of neutrons $n(t)$ can be modeled by some kind of "function" where the input variable is reactivity $\rho(t)$. Then, by considering the inverse function as the input variable of $n(t)$, can we deduce the reactivity from the measured neutron count rates? The "inverse kinetics method" is a technique to estimate the time variation of the reactivity $\rho(t)$ using the measured time variation of $n(t)$.

First, let us suppose that an external neutron source of intensity S is inserted into a core where the reactivity is $\rho(t)$. Even if the reactivity $\rho(t)$ changes with time due to the control rod operation, the numbers of neutrons and delayed neutron precursors, $n(t)$ and $C(t)$, respectively, can be described based on the point kinetics equation in a similar way as the source jerk method:

$$\frac{dn(t)}{dt} = \frac{\rho(t) - \beta}{\Lambda} n(t) + \lambda C(t) + S, \tag{2.38}$$

$$\frac{dC(t)}{dt} = \frac{\beta}{\Lambda} n(t) - \lambda C(t). \tag{2.39}$$

If the time variation of reactivity $\rho(t)$ can be given as the input variable for Eqs. (2.38) and (2.39), the time variation of the neutron number $n(t)$ can be numerically predicted using the numerical integration of the first-order differential equation, e.g., the Euler or Runge–Kutta method.

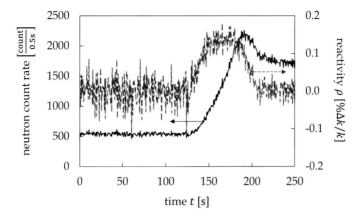

Fig. 2.9 Example of reactivity estimation using inverse kinetics method

Reversing this idea, we can numerically estimate the time variation (increase or decrease) of $\rho(t)$ as shown in Fig. 2.9, by giving $n(t)$ as the input variable. This method is called the "inverse kinetics method," and the measurement principle for the "reactivity meter" that is widely utilized for the real-time monitoring of reactivity in nuclear reactors [9].

As a precondition, let us assume that the neutron count rate $n(t)$ is proportional to the total number of neutrons in the core and the time-series data are successively measured with a certain time width Δt using a neutron detector located in the core. For simplicity, by considering delayed neutrons group with one group, the number of delayed neutron precursors $C(t)$ can be numerically estimated by the following equation using the input value of the measured neutron count rate $n(t)$:

$$C(t) \approx C(t - \Delta t)e^{-\lambda \Delta t} + \frac{1 - e^{-\lambda \Delta t}}{\lambda} \frac{\beta}{\Lambda} n(t). \tag{2.40}$$

Equation (2.40) can be derived based on Eq. (2.39), and $C(t - \Delta t)$ denotes the number of delayed neutron precursors at the previous time $t - \Delta t$. Note that the initial value of $C(0) = C_0$ can be estimated by Eq. (2.34) if the core is a steady state at time $t = 0$. By giving C_0 as the initial value for Eq. (2.40), $C(t)$ can be successively estimated from the continuous measurements of $n(t)$. Finally, the reactivity $\rho(t)$ can be inversely estimated by substituting the estimated values of $C(t)$ into the following equation that is obtained by transforming the one-point kinetics equation of Eq. (2.38):

$$\rho(t) = \frac{\Lambda}{n(t)} \left(\frac{dn(t)}{dt} + \frac{dC(t)}{dt} - S \right)$$

$$\approx \frac{\Lambda}{n(t)} \left(\frac{n(t) - n(t - \Delta t)}{\Delta t} + \frac{C(t) - C(t - \Delta t)}{\Delta t} - S \right). \tag{2.41}$$

Equation (2.41) is simplified by considering only delayed neutrons with one group. If delayed neutrons with six groups are considered, the reactivity $\rho(t)$ can be more rigorously estimated using the modified equation as follows:

$$
\rho(t) = \frac{\Lambda}{n(t)} \left(\frac{dn(t)}{dt} + \sum_{i=1}^{6} \frac{dC_i(t)}{dt} - S \right)
$$

$$
\approx \frac{\Lambda}{n(t)} \left\{ \frac{n(t) - n(t - \Delta t)}{\Delta t} + \left(\sum_{i=1}^{6} \frac{C_i(t) - C_i(t - \Delta t)}{\Delta t} \right) - S \right\}, \quad (2.42)
$$

where the numerical value of the delayed neutron precursor, $C_i(t)$, can be obtained by Eq. (2.40) with the ith delayed neutron fraction $\beta_{\mathrm{eff},i}$ and decay constant λ_i summarized in Table 1.4.

In order to inversely estimate the reactivity $\rho(t)$ using the inverse kinetics method of Eq. (2.42), it is necessary to provide not only the point kinetics parameters but also the value of the source intensity S (neutrons/s). If the core is in a critical state without an external neutron source, the source intensity is given by $S = 0$. On the other hand, in the case of a source-driven subcritical core, if the initial subcriticality $(-\rho_0)$ is known in advance and the initial neutron count rate n_0 in the steady state can be measured, the effective intensity S can be calibrated as follows:

$$
S = -\rho_0 \frac{n_0}{\Lambda}. \tag{2.43}
$$

2.3.3.2   Method of Measurement

To measure the reactivity using the inverse kinetics method for a critical core or a source-driven subcritical core, the experimental procedures are described as follows:

(1) In advance, the excess reactivity and the control rod worth are measured by the positive period method and the rod drop method, respectively.
(2) If the initial condition is a critical state without an external neutron source, the control rod positions are adjusted to maintain an appropriate constant value of neutron count rate. In this case, the following procedures (3) to (5) are skipped.
(3) If the initial condition is a steady state of a subcritical core with an external neutron source, four control rods (two safety rods: SRs, a shim safety rod: SSR, and a regulating rod: RR) are arbitrarily inserted to configure a certain subcritical system. Then, let us wait until the neutron count rate becomes a constant value.
(4) For example, the measurements of the neutron count during 100 s are repeated five times. From the five measured counts, the sample mean value of neutron count rate n_0 is calculated.
(5) Using the excess reactivity and control rod worth, the initial subcriticality $(-\rho_0)$ is obtained. Then, the effective source intensity S is determined by Eq. (2.43).

(6) A neutron measurement system is constructed to successively measure the neutron count rate $n(t)$ at a certain time interval Δt. Typically, $\Delta t = 0.5$ s.

(7) While the time-series data of neutron count rate are successively measured, the reactivity is arbitrarily changed by adjusting the control rod position. In this operation, the time variation of the control rod position should be recorded.

(8) Based on the inverse kinetics method, Eqs. (2.40) and (2.42), with the measured neutron count rate $n(t)$, the time variation of reactivity $\rho(t)$ is estimated.

2.3.3.3 Discussion

Let us discuss the experimental results of reactivity $\rho(t)$ using the inverse kinetics method from the following viewpoints:

(1) Let us investigate the relationship between the time variations of the neutron count rate $n(t)$ (e.g., magnitude and sign for the slope) and the estimated reactivity $\rho(t)$ by the inverse kinetics method. For example, how the reactivity $\rho(t)$ is inversely estimated for the following three cases of $n(t)$?

(a) $n(t)$ is constant
(b) $n(t)$ increases with time
(c) $n(t)$ decreases with time.

(2) As can be seen from the experimental results of $\rho(t)$ using the inverse kinetics method, the estimated $\rho(t)$ has statistical uncertainty. Why does the statistical uncertainty of $\rho(t)$ arise? Let us consider how the statistical uncertainty of $\rho(t)$ can be reduced in the inverse kinetics method.

(3) Let us evaluate the reference value of reactivity ρ_{ref} at a specific time t, based on the records of control rod positions, the excess reactivity ρ_{excess}, and the control rod worth $\rho_{rod,i}$. By comparing the reactivity estimated by the inverse kinetics method with the reference value ρ_{ref}, is there a significant difference in ρ between them? If any, what is the major reason?

(4) Are there any differences in the reactivity $\rho(t)$ between the neutron source multiplication and inverse kinetics methods? Let us compare and discuss these results by both methods for the following cases: (a) during the control rod operation, and (b) after sufficient time has passed from the end of control rod operation.

2.3.4 Reactor Noise Analysis Method

2.3.4.1 Principle of Measurement

Let us suppose that the time-series data of neutron counts are successively measured in a steady state for a source-driven subcritical core or a critical core with an external neutron source. Then, as shown in Fig. 2.10, the measured neutron counts fluctuate around a certain mean value. As explained later, there is a unique phenomenon that the "reactor noise" (fluctuation of the neutron count from the mean value) becomes larger as the target core approaches the criticality or the subcriticality becomes shallower. Based on this phenomenon, the subcriticality measurement technique by analyzing the measured reactor noise is called the "reactor noise analysis method."

First, let us consider the phenomenon of the decays of radioactive nuclides. The probability of the radioactive decay per unit time is constant, and these decays are independent and random events each other. Therefore, it is well known that the number of decays of radioactive nuclides (or the total number of detected primary radiation emitted by the decay), C, follows the Poisson distribution. According to the statistical property of the Poisson distribution, the standard deviation (i.e., a measure for the amount of dispersion from the mean value) can be approximated by $\sqrt{C}$.

Next, let us consider the number of neutrons in a nuclear reactor where fission chain reactions happen. In this case, neutrons emitted by an external neutron source or delayed neutron precursor have the role of "seed" to yield the fission chain of "neutron family tree." Because of the fission chain reaction, the neutron-descendants belonging to the same fission chain family form a group like "fish shoal." Let us

Fig. 2.10 Examples of temporal fluctuations in neutron counts (reactor noises)

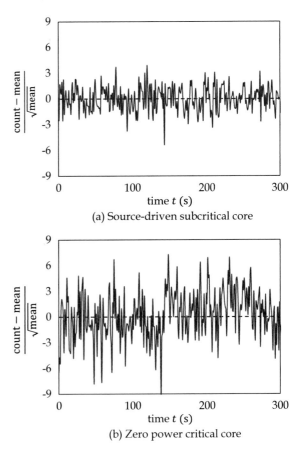

(a) Source-driven subcritical core

(b) Zero power critical core

suppose that, like the fishing for the shoal, some neutrons in the group are detected using a neutron detector. If the neutron multiplication factor is $k_{\text{eff}} = 0$ (i.e., the neutron family disappears immediately after the first generation), we have to wait for the next neutron emitted by the neutron source after one neutron is detected, because there are no neutron family within the neighborhood of the first detected neutron. On the other hand, if the neutron multiplication factor k_{eff} is close to 1 to induce the longer fission chain reaction, we expect a higher possibility that other neutrons belonging to the same fission chain can be detected within the neighborhood of the first detected neutron. In other words, if a neutron were a fish, we have a better chance to catch other descendant neutrons by dropping a fishing rod immediately after catching one neutron. Therefore, as shown in Fig. 2.10, there are sparse or dense time domains in the reactor noise like "fish shoal".

Various reactor noise analysis methods have been proposed in the field of reactor physics experiments [11]. Among them, the "Feynman-α method (variance-to-mean ratio method) [10]" is explained below in this section. Let us suppose that the neutron counts $C(T)$ are detected during a counting gate of width T in a steady state for a target subcritical core, as shown in Fig. 2.11. By successively measuring the time-series data of neutron counts $(C_1(T), C_2(T), \cdots, C_N(T))$, where N is the total number of count data, the sample mean $\mu(T)$ and the unbiased variance $\sigma^2(T)$ are calculated as follows:

$$\mu(T) = \frac{1}{N} \sum_{i=1}^{N} C_i(T), \tag{2.44}$$

$$\sigma^2(T) = \frac{1}{N-1} \sum_{i=1}^{N} (C_i(T) - \mu(T))^2. \tag{2.45}$$

If no fission chain reaction occurs in the target system, neutrons are randomly detected. Then, the frequency distribution of $C(T)$ follows the Poisson distribution,

Fig. 2.11 Example of reactor noise analysis using the Feynman-α method

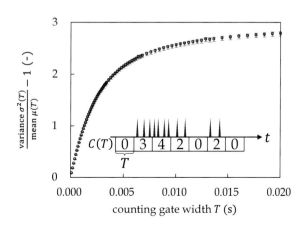

thus the mean $\mu(T)$ is equal to the variance $\sigma^2(T)$. On the other hand, as more fission chain reactions occur due to the shallower subcriticality, the variance $\sigma^2(T)$ becomes larger than the mean $\mu(T)$ because of larger differences between the sparse and dense time domains in the measured $C(T)$. By focusing on this feature, the magnitude of "neutron correlation" caused by the fission chain reaction can be investigated by analyzing the relative deviation of variance in neutron counts compared with the variance in the case of the Poisson distribution. Namely, the neutron correlation Y value is defined as follows:

$$Y(T) = \frac{\sigma^2(T)}{\mu(T)} - 1. \tag{2.46}$$

Now, let us analyze the variation of $Y(T)$ defined by Eq. (2.46) for various counting gate widths T. For example, as shown in Fig. 2.12, if the time-series data of $C(T)$ are continuously measured, different Y values (e.g., $Y(2T)$ and $Y(4T)$) corresponding to the gate widths $2\ T$ and $4\ T$) by bunching the adjacent data of neutron counts (e.g., $C_i(T)$ and $C_{i+1}(T)$). As shown in Fig. 2.11, $Y(T)$ is almost equal to zero at $T = 0$. Namely, when the counting gate width T is very small, the measured neutron counts can be reasonably regarded as the random detection process which follows the Poisson distribution. On the other hand, as T becomes larger, $Y(T)$ converges to a certain saturation value. Although the detailed theoretical derivation of the Feynman-α method is omitted in this section, the variation of $Y(T)$ can be expressed by the following analytical formula [12]:

$$Y(T) = Y_\infty \left(1 - \frac{1 - e^{-\alpha T}}{\alpha T} \right) + c_1 + c_2 T \approx Y_\infty \left(1 - \frac{1 - e^{-\alpha T}}{\alpha T} \right), \tag{2.47}$$

$$Y_\infty = \varepsilon \frac{\langle \nu(\nu - 1) \rangle}{\langle \nu \rangle^2} \frac{1}{(\beta_{\text{eff}} - \rho)^2}. \tag{2.48}$$

In Eq. (2.47), Y_∞ and α represent the saturation value of Y and the prompt neutron decay constant (s^{-1}), respectively. Furthermore, c_1 and c_2 are additional terms to correct effects on the Y value mainly due to the dead time and the delayed neutrons, respectively. If these effects are negligible, we can approximate Eq. (2.47) by $c_1 = c_2 = 0$. In Eq. (2.48), ε is the detection efficiency, ν is the number of fission neutrons

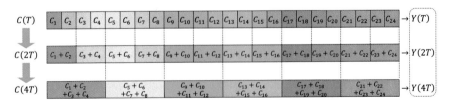

Fig. 2.12 Bunching process for time-series data of neutron counts

emitted per fission reaction, the bracket $\langle\,\rangle$ is the expected value, and $\langle \nu(\nu - 1)\rangle / \langle \nu \rangle^2$ is the nuclear data called the "Diven factor" [13].

If the position of the neutron detector and the external neutron source in the target core is the same and only the subcriticality $(-\rho)$ changes, Eq. (2.48) indicates that the saturation value Y_∞ increases inversely proportional to $(\beta_{\text{eff}} - \rho)^2$ (i.e., the square of the negative reactivity considering only prompt neutrons) as the subcriticality becomes shallower. However, the subcriticality is not easily estimated only from the saturation value Y_∞, because Y_∞ is also proportional to the detection efficiency ε and some kind of measurement is necessary to determine ε. Therefore, in the Feynman-α method, the prompt neutron decay constant α is often estimated using the nonlinear least-squares fitting method using Eq. (2.47) for the measured $Y(T)$ values as shown in Fig. 2.11. Here, the prompt neutron decay constant α is a time constant (s^{-1}) that expresses how fast the neutron family decreases exponentially. The relationship between α and the subcriticality $(-\rho)$ can be well approximated as follows:

$$\alpha = \frac{\beta_{\text{eff}} - \rho}{\Lambda} = \frac{1 - (1 - \beta_{\text{eff}})k_{\text{eff}}}{\ell}. \tag{2.49}$$

Using Eq. (2.49) with the point kinetics parameters Λ and β_{eff} for the UTR-KINKI, the subcriticality $(-\rho)$ can be indirectly estimated from the measurement results of prompt neutron decay constant α.

2.3.4.2 Method of Measurement

To measure the prompt neutron decay constant α using the Feynman-α method for a target subcritical core, the experimental procedures are described as follows:

(1) To measure the reactor noise, a neutron detector (e.g., BF$_3$ detector) is placed in a hole in the graphite reflector, and a neutron measurement system is constructed as shown in Fig. 2.2.

(2) Four control rods (two SRs, SSR and RR) are arbitrarily inserted to configure a certain subcritical system. Or, a shutdown state (i.e., all control rods fully inserted) may be useful to carry out a longer reactor noise measurement using the inherent neutron source in the nuclear fuel.

(3) If an external neutron source is inserted into the core, or even if the inherent neutron source due to the U$(\alpha, n)^{27}$Al reaction is utilized, let us wait until the neutron count rate becomes a constant value.

(4) According to the subcriticality, the counting gate of width T should be appropriately set to analyze the prompt neutron decay constant α. Typically, $T = 0.1$ ms $= 10^{-4}$ s. The time-series data of neutron count $(C_1(T), C_2(T), \cdots, C_N(T))$ are successively measured using the basic gate width T.

(5) From the measured reactor noise, the Feynman-α histogram is analyzed using the bunching method to evaluate the prompt neutron decay constant α.

(6) Based on Eq. (2.49), the subcriticality $(-\rho)$ is finally converted from the measured α value.

2.3.4.3 Discussion

Let us discuss the experimental results of Feynman-α method from the following viewpoints:

(1) Let us evaluate the reference value of subcriticality $(-\rho_{ref})$ based on the control rod pattern using the excess reactivity ρ_{excess} and the control rod worth $\rho_{rod,i}$ in advance. By comparing the subcriticality estimated by the Feynman-α method with the reference value $(-\rho_{ref})$, is there a significant difference in $(-\rho)$ between them? If any, what is the major reason?

(2) Let us suppose that the subcriticality of a target core can be varied by changing the control rod pattern. According to the magnitude of subcriticality $(-\rho)$, how do the saturation value Y_{∞} and the prompt neutron decay constant α change? Let us discuss the physical meanings of the above-mentioned changes.

(3) In the Feynman-α method, because of the statistical uncertainty of neutron counts, the Y and α values also have statistical uncertainties. Let us discuss how we can reduce the statistical uncertainties of the Y and α values.

[Column] Statistical uncertainty in the Feynman-α method

Because the total measurement time is limited in the actual reactor noise measurement, it is a complicated question to quantify the statistical uncertainty σ_Y in the measured Y value (i.e., the uncertainty of fluctuation in neutron counts). Various estimation methods for the statistical uncertainty σ_Y have been studied, e.g., (1) the resampling technique using the moving block bootstrap method and (2) the analytical formula using the uncertainty propagation (or sandwich formula). The latter method (2) utilizes the statistical property that the probability distribution of neutron counts in a neutron multiplication system follows a special distribution known as the Pál-Mogil'ner-Zolotukhin-Bell-Babala (PMZBB) distribution [14, 15]. For example, if the subcriticality for the target core is $(-\rho) < 10,000$ pcm $(= 10\% \Delta k/k = 0.1$ $\Delta k/k)$, the statistical uncertainty σ_Y can be easily estimated by the following formula using the total number of count data N, the sample mean μ, and the measured Y value [16]:

$$\sigma_Y \approx (Y+1)\sqrt{\frac{Y(2Y+1)(5Y+2)}{N(Y+1)^2\mu} + \frac{2}{N-1}}. \tag{2.50}$$

References

1. Duderstadt JJ, Hamilton LJ (1976) Nuclear reactor analysis. Wiley, New York
2. Misawa T, Unesaki H, Pyeon CH (2010) Nuclear reactor physics experiments. Kyoto University Press, Kyoto, Japan. http://hdl.handle.net/2433/276400. Accessed 23 Oct 2022
3. Atomic Energy Society of Japan, Reactor Physics Division ed (2008) Genshiro Butsuri (Reactor physics) (textbook of reactor physics: intermediate edition). Atomic Energy Society of Japan, Tokyo, Japan (in Japanese). https://rpg.jaea.go.jp/else/rpd/others/study/text_aesj.html. Accessed 1 July 2022
4. Nagaya Y, Okumura K, Mori T (2015) Recent development of JAEA's Monte Carlo code MVP for reactor physics applications. Ann Nucl Energy 82:85–89
5. Shibata K, Iwamoto O, Nakagawa T et al. JENDL-4.0: a new library for nuclear science and engineering. J Nucl Sci Technol 48:1–30
6. Hogan WS. Negative-reactivity measurements. Nucl Sci Eng 8:518–522
7. Sakon A, Nakajima K, Hohara S et al (2019) Experimental study of neutron counting in a zero-power reactor driven by a neutron source inherent in highly enriched uranium fuels. J Nucl Sci Technol 56:254–259
8. Endo T, Nonaka A, Imai S et al (2020) Subcriticality measurement using time-domain decomposition-based integral method for simultaneous reactivity and source changes. J Nucl Sci Technol 57:607–616
9. Sastre CA (1960) The measurement of reactivity. Nucl Sci Eng 8:443–447
10. Feynman RP, de Hoffmann F, Serber R (1956) Dispersion of the neutron emission in U-235 fission. J Nucl Energy 3:64–69
11. Williams MMR (1974) Random processes in nuclear reactors. Pergamon Press, Oxford, UK
12. Hashimoto K, Mouri T, Ohtani N (1999) Reduction of delayed-neutron contribution to variance-to-mean ratio by application of difference filter technique. J Nucl Sci Technol 36:555–559
13. Diven BC, Martin HC, Taschek RF et al (1956) Multiplicities of fission neutrons. Phys Rev 101:1012–1015
14. Saito K (1979) Source papers in reactor noise. Prog Nucl Energy 3:157–218
15. Endo T, Nishioka F, Yamamoto A et al (2022) Theoretical derivation of a unique combination number hidden in the higher-order neutron correlation factors using the Pál-Bell equation. Nucl Sci Eng (in print). https://doi.org/10.1080/00295639.2022.2049992
16. Endo T, Yamamoto A (2019) Comparison of theoretical formulae and bootstrap method for statistical error estimation of Feynman-α method. Ann Nucl Energy 124:606–615

Chapter 3
Radiation Measurement and Application

Abstract Unlike a power reactor, a research reactor is mainly used as a neutron generator for various experiments and applications. The UTR-KINKI is a low-power research reactor installed for the purpose of research and education at the university. Although its rated thermal power is only 1 W, it can be used as a neutron source for various educational programs on radiation measurement and applications. In addition, when conducting experiments, the γ-rays and neutrons emitted from the reactor must be measured properly from the viewpoint of radiation protection. This chapter describes experimental programs using a research reactor as a neutron source and the measurements necessary for radiation control during the experiment.

Keywords Activation reaction · Thermal neutron flux · Neutron dose rate · Gamma dose rate · Gamma spectrometry · X-ray radiography · Neutron radiography

3.1 Activation and Half-Life Measurement

3.1.1 Activation

Aluminum (Al) is a stable isotope with a mass number of 27, which has a natural abundance ratio of 100%. When a small piece of Al is placed in a nuclear reactor and irradiated with thermal neutrons, a part of the nuclide Al-27 is induced by capture reactions shown in the following reaction formula, changing into Al-28:

$$^{27}\text{Al} + {}^{1}n \rightarrow {}^{28}\text{Al}.$$

The above reaction formula can be expressed as $^{27}\text{Al}(n, \gamma)^{28}\text{Al}$. The reaction cross sections of Al are shown in Fig. 3.1. As shown in Fig. 3.1, in the case of Al-27 as the target nucleus, manganese-27 (Mg-27), sodium-24 (Na-24) and Mg-26 are produced by the (n, p), (n, α) and (n, np) reactions, respectively, in addition to the (n, γ) reaction in the above formula. In terms of reactor irradiation, the (n, γ) reaction, which is the "capture reaction" in Fig. 3.1, occurs mainly because the thermal neutron component

© The Author(s) 2023
G. Wakabayashi et al., *Introduction to Nuclear Reactor Experiments*,
https://doi.org/10.1007/978-981-19-6589-0_3

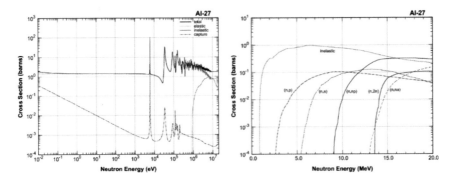

Fig. 3.1 Cross sections of Al-27 for neutrons (Data from Ref. [1])

is dominant. The thermal neutron capture cross section is relatively large, 0.23 barn. The unit of the cross section is barn, 10^{-24} cm ^{2}.

The amount of Al-28 produced is proportional to the following four factors: the amount of the target nuclide (number of atoms); the particle flux density of thermal neutrons ϕ (cm^{-2} s^{-1}) (this is called the thermal neutron flux); the probability of the nuclear reaction occurring σ (10^{-24} cm^2); the irradiation time t (s). If the irradiation position, irradiation time and target nuclide are the same, the produced radioactivity (Bq) increases with the irradiated sample amount (mg).

The Al-28 is the radionuclide that decays to silicon-28 (Si-28; stable isotope) with a half-life of 2.24 min. At the time of decay, β^--rays with a maximum energy of 2.86 MeV are emitted, followed by γ-rays with a maximum energy of 1.779 MeV (Ref. [2]). Figure 3.2 shows the decay diagram. The produced radioactivity can be quantified by measuring these radiations with a radiation instrument. Based on the produced radioactivity, the number of atoms in the target sample can be obtained from the thermal neutron flux ϕ, reaction cross section σ and irradiation time t.

The analysis method to determine the amount of contained elements in a sample is called the activation analysis, using "activation" to convert stable elements into radioactive isotopes. Since neutrons have no electric charge, they are absorbed by

Fig. 3.2 Diagram of the decay of Al-28

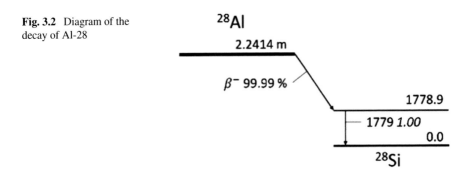

atomic nuclei without being subjected to Coulomb repulsive force, causing the formation of neutron-rich nuclei. Therefore, it is relatively easy to cause activation of the
nucleus. Since a nuclear reactor is the device that provides neutrons stably, activation
analysis can be conducted by using a nuclear reactor.

The activation analysis method is one of the elemental analysis methods. Because
of its high sensitivity, the activation analysis method can be applied as an ultra-trace
analysis method. Also, the activation analysis method has excellent characteristics:
(1) high-analytical sensitivity; (2) non-destructive analysis without chemical treatment; (3) simultaneous analysis of many elements (Ref. [3]). Furthermore, the activation analysis method is used in a wide range of fields, including earth science,
biology and archaeology, for the determination of trace elements (Na, cobalt: Co,
copper: Cu, etc.) in rocks, minerals, natural water, meteorites and blood, and for
the analysis of trace elements (gold: Au, Cu, mercury: Hg, etc.) in hair. To obtain
high sensitivity, it is necessary that the cross section of the target nuclear reaction is
large, the half-life of the produced radionuclide is suitably short, and the irradiation
neutron flux is large.

Another application of "activation," besides the activation analysis, is the production of Si semiconductor by neutron irradiation. An n-type Si semiconductor is the
Si single crystal doped with phosphorus (P) (Ref. [4]). The Si-30 is naturally present
in Si at a ratio of 3.05%. When the Si single crystal is irradiated with neutrons, Si-30
absorbs neutrons and changes into Si-31. The Si-31 is induced by β-decay and is
converted to the P-31 that is a stable isotope. This reaction can be used to dope P
uniformly in Si. By controlling the neutron irradiation time, the concentration of P
to be added can be set precisely.

3.1.2 Decay and Half-Life of Radioactivity

The phenomenon that a radioactive isotope changes into a stable element is called
"decay," occurring stochastically. Radioactive nuclei (radionuclides) have their own
half-lives, and the number of radioactive nuclei decreases with time. The Al-28 is
the radionuclide that decays to Si-28 with a half-life of 2.24 min. Upon the decay,
β-rays with a maximum energy of 2.86 MeV are emitted, which can be measured
with a GM (Geiger–Mueller) counter.

Because radiation is emitted in accordance with a certain of decay, the number
of decays per second is associated with the ability to emit radiation and is called the
radioactivity, which is written in the following equation:

$$A = -\frac{dN}{dt} = \lambda \cdot N, \tag{3.1}$$

where A is the radioactivity (Bq), N the number of atoms in the target material
(number of atoms), and λ the decay constant (s^{-1}). As shown in Eq. (3.1), the radioactivity is proportional to the number of target atoms, and when the above differential

Fig. 3.3 Decay of
radioactivity

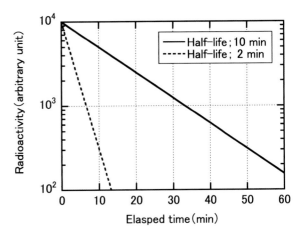

equation is solved, the following equation is obtained, and the radioactivity decreases
according to an exponential function if there is no supply of target atoms from other
sources:

$$A(t) = A_0 e^{-\lambda t}. \tag{3.2}$$

Here, we assume that the radioactivity A is the value of A_0 at time $t = 0$. A semilogarithmic graph with the elapsed time of decay of the radioactivity as the horizontal
axis is shown in Fig. 3.3. The half-life $T_{1/2}$ and the decay constant are given by

$$T_{1/2} = \frac{\ln 2}{\lambda}. \tag{3.3}$$

From Eq. (3.3), the decay constant λ is inversely proportional to the half-life $T_{1/2}$.
If the decay constant is large, i.e., the half-life is short, the slope of the straight line
is large, as shown in Fig. 3.3.

3.1.3 GM Counter Tube

The Al-28 that is produced by neutron irradiation in a nuclear reactor is a β^--decay
nuclide. The β^--rays emitted by the decay are counted by GM counters. The counting
rate $(s^{-1}$; cps: counts per second) corresponds to the number of radiation emitted per
unit time. Also, number of radiation corresponds to the radioactivity of the sample
to be measured. The half-life is obtained from the decrease of the counting rate in
accordance with time.

The GM counter is a kind of gas ionization detector in which a gas is enclosed.
Also, the GM counter is generally designed with a hollow cylinder as the negative
electrode and a thin wire-shaped positive electrode in the center of the cylinder. The

Fig. 3.4 Photograph of GM
counter and scaler

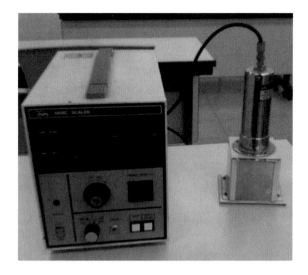

gas is ionized by the radiation incident in the tube, and the generated electrons and cations move toward the positive and negative electrodes, respectively. Low-energy electrons are also accelerated by the electric field strength, and an electron avalanche is generated. About 10^9 times as much charge is generated in the tube as is initially generated by the radiation, and the pulse output from the counting tube can be a few V or more, which makes it highly resistant to noise and provides high detection sensitivity (Ref. [5]).

The GM counter used in this experiment is an end-window-type GM counter, which has a mica window at the tip of the cylindrical probe. The pulse signal from the GM counter is introduced to the scaler for counting. The GM counter and the scaler used in the experiment are shown in Fig. 3.4 for reference. The scaler module shown in the picture has also the function of applying a high voltage to the counter tube.

3.1.4 Method of Measurement

3.1.4.1 Preparation of GM Counter 1: Plateau Curve

GM counter tubes are used in the following procedures:

(1) Connect the GM counter to the scaler with a coaxial cable.
(2) Check that the voltage adjustment volume of the high-voltage power supply indicates the lowest voltage.
(3) Turn on the power switch, and increase slowly and gradually the applied voltage to fix it at the specified voltage.
(4) Set the timer and start counting.

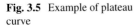
Fig. 3.5 Example of plateau
curve

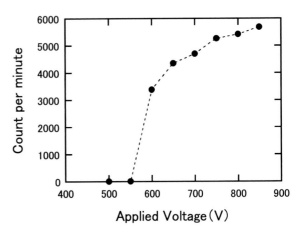

Applied Voltage (V)

The voltage to be used is determined by obtaining the plateau curve of the GM counter while changing the applied voltage using a β-ray radiation source. Figure 3.5 shows an example of the plateau curve. The working voltage should be set at about 1/3 of the plateau length from the counting start voltage, which is 700 V, as shown Fig. 3.5.

3.1.4.2 Preparation of GM Counter 2: Background Counting

When measuring radiation with a GM counter, background radiation emitted from natural radionuclides in the vicinity is counted in addition to the radiation emitted from the measurement target. When the counting rate decreases due to the decay of radioactivity, the decrease in the counting rate is slowed down by the addition of counting due to the background radiation. To obtain a net count of the radiation emitted from the object to be measured, it is necessary to measure the counting rate due to background radiation in advance. The counting rate due to background radiation varies depending on day and night, season, weather, etc., even at the same measurement site. Also, in most cases, the variation can be ignored if the measurement is made within one hour.

Before conducting the measurement of the induced radioactivity, the background count rate is measured by the following procedures:

(1) Connect the GM counter and the scaler with a coaxial scale, and turn on the power.
(2) Without inserting a sample, set the applied voltage to the working voltage.
(3) Set the preset time to 10 min, and measure the background count. Record the results.

3.1.4.3   Experimental Procedures

A small piece of Al is placed in the reactor at the full power of 1 W, and thermal neutron irradiation is carried out, ranging between 15 and 30 min.

A stopwatch is operated at the same time as the end of irradiation, and counting is started as soon as possible after the end of irradiation. The GM counter is used to count the β-rays emitted from the Al piece for 1 min each. The procedure is repeated at one-minute intervals. In each counting, the counting start time (elapsed time after the end of irradiation) and the counting rate (counts per minute: cpm) are recorded. An example of the recording format is shown in Table 3.1.

The specific procedure of the decay measurement of radioactivity (Al) is described as follows:

(1)   Connect the GM counter and the scaler, and turn on the power.
(2)   Set the applied voltage to the operating voltage.
(3)   Set the preset time to one minute.
(4)   Set a small piece of Al to be measured on the sample table.
(5)   Press the "RESET COUNT" button according to the elapsed time since the end of irradiation to start counting.
(6)   The start time of counting (display time of the stopwatch) is recorded, and the system waits until the counting is completed.
(7)   After completing the measurement by presetting, record the count (Table 3.1).
(8)   Repeat steps (5) to (7) until 10 half-lives (about 23 min) have elapsed from the end of irradiation.
(9)   The net counting rate is calculated by subtracting the background counting rate from the counting rate.
(10) As shown in Fig. 3.6, the vertical axis of the semilogarithmic graph sheet is plotted as the net counting rate (cpm), and the horizontal axis as the elapsed

Table 3.1 Datasheet of decay measurement of irradiated sample

Date of experiment	Year/month/day	2020/06/10
Irradiation start time	Hour/minute/second	11:05:00
Irradiation end time	Hour/minute/second	11:20:00
Background count rate	Count per second	30.0 (cpm)
Elapsed time (min)	Counting rate (cpm)	Net counting rate (cpm)
5	10,030	10,000
7	5418	5388
9	2933	2903
11	1594	1564
13	873	843
15	484	454
17	275	245

Fig. 3.6 Graph sheet of
elapsed time and net
counting rate

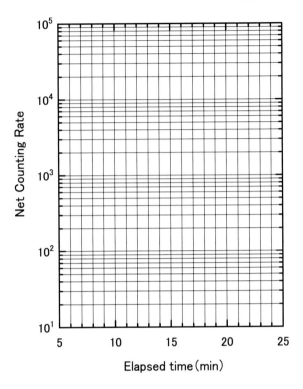

time (min) after the end of irradiation. Since the radioactivity decreases exponentially with time, the decay curve is a straight line with a decreasing right
side on the semilogarithmic graph. From the decay curve, the time required for
the counting rate to decrease to half is obtained with 3 to 4 significant digits
and is used as the half-life of Al-28.

3.1.5   Discussion

The key points for discussion are as follows:

(1) The half-life of Al-28 was determined by obtaining the time required for the
counting rate to be halved from the decay curve. Verify that the value is close
to the literature value of 2.2414 min (134.48 s).
(2) The half-life you obtained has an error. Discuss the methods used to minimize
the error.
(3) GM counter tubes have a dead time from 10^{-4} s to twice that. Therefore, when
the counting rate is 300 per second (cps), the counting error is about 0.3%. In
addition, when the counting rate is equal to the background counting rate, the
counting error of the net counting rate is large. In drawing the decay curve,

consider the range of net counting rates that are appropriate for obtaining an accurate half-life.

(4) Calculate the radioactivity after the end of irradiation, assuming that the absolute counting efficiency of the GM counter (number counted/number of emitted beta-rays) is 0.15. Also, calculate the number of Al-28 atoms in the calculated radioactivity.

3.2 Thermal Neutron Flux Measurement

3.2.1 Introduction

A variety of neutron irradiation experiments is performed by using research reactors. Samples and detectors are placed inside or outside the reactor and irradiated with neutrons to analyze the composition of the samples or to investigate the response of the detectors. In such experiments, information on the neutron flux at the location where the sample or detector is irradiated is essential to determine the neutron fluence as the time-integrated amount of neutron flux, which must be measured accurately for each irradiation. In addition, the spatial distribution of the neutron flux inside the reactor is determined by the shape, structure and composition of the reactor and the position of the control rods, and is proportional to the thermal power distribution in the reactor. Therefore, we can also know the accurate thermal power distribution inside the reactor by measuring the spatial distribution of the neutron flux. Especially, in an extremely low power reactor such as UTR-KINKI, the thermal power cannot be determined from the temperature change of the reactor because the reactor is kept at normal temperature and pressure during operation. That is why the thermal power of the reactor is obtained from the spatial distribution of neutron flux.

The aim of this experiment is to measure a thermal neutron flux in the reactor by the activation method, which is the most widely used method for measuring neutron flux, and to understand the principle of the measurement.

3.2.2 Theory

When a material consisting of a stable isotope is irradiated with neutrons, activation reactions may occur such that the stable isotope absorbs neutrons and becomes a radioactive isotope. The activation method determines the flux of irradiated neutrons indirectly by measuring the activity produced in the material.

Let us consider the relationship between the neutron flux and the activity produced in the target material. Assuming that the neutron flux is ϕ ($cm^{-2} s^{-1}$), the cross section of the activation reaction is σ (cm^2), and the number of target nuclei is N, the reaction rate (the number of activation reactions per unit time) R (s^{-1}) is expressed by the following equation:

$$R = N\sigma\phi. \tag{3.4}$$

The produced radioactive nuclei will decay with its decay constant. The rate of decay (the number of decays per unit time) is expressed as λn (s^{-1}), where λ is the decay constant (s^{-1}), and n is the number of radioactive nuclei in the target material. Therefore, the rate of change in n can be expressed as follows:

$$\frac{dn}{dt} = R - \lambda n. \tag{3.5}$$

Assuming $n = 0$ at $t = 0$, and solving the differential equation for n, we get the solution as follows:

$$n = \frac{R}{\lambda}\left(1 - e^{-\lambda t}\right). \tag{3.6}$$

Since the activity A of the material is given by λn, the following equation is obtained by multiplying both sides of Eq. (3.6) by the decay constant λ:

$$A = R\left(1 - e^{-\lambda t}\right). \tag{3.7}$$

This is the activity of the target material produced for a time t after the start of the irradiation.

As the time t approaches to infinity ($t \to \infty$) in Eq. (3.7), the term $\left(1 - e^{-\lambda t}\right)$ is unity and the activity approaches an asymptote. The activity A_∞ is called the saturation activity, and the following relationship can be obtained:

$$A_\infty = R. \tag{3.8}$$

Here, as shown in Fig. 3.7, consider that after the irradiation from $t = 0$ to $t = t_0$, the preparation for the activity measurement is made from $t = t_0$ to $t = t_1$, and the measurement is performed from $t = t_1$ to $t = t_2$. Since the activity decays exponentially after the irradiation is stopped at $t = t_0$, if the activity at the end of irradiation ($t = t_0$) is A_0, the activity of the target material A is given by

$$A = A_0 e^{-\lambda(t - t_0)}$$
$$= R\left(1 - e^{-\lambda t_0}\right)e^{-\lambda(t - t_0)}. \tag{3.9}$$

Therefore, the total count C obtained from the measurement performed for the interval between t_1 and t_2 can be expressed as follows:

$$C = \varepsilon \int_{t_1}^{t_2} A_0 e^{-\lambda(t - t_0)} dt + B$$
$$= \frac{\varepsilon A_0}{\lambda}\left(e^{-\lambda t_1} - e^{-\lambda t_2}\right) + B, \tag{3.10}$$

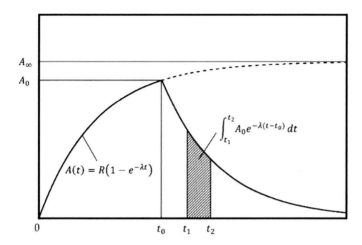

Fig. 3.7 Change in radioactivity of target material irradiated with neutrons for time t_0

where B is the background count and ε is the detection efficiency. The reaction rate R (or saturated activity) can be obtained by measuring C and B, as follows:

$$R = A_\infty = \frac{\lambda(C - B)}{\varepsilon\left(1 - e^{-\lambda t_0}\right)\left(e^{-\lambda t_1} - e^{-\lambda t_2}\right)}. \tag{3.11}$$

Once the reaction rate R is obtained in this way, the neutron flux ϕ can be obtained from the relationship between the reaction rate and the neutron flux shown in Eq. (3.4) as follows:

$$\phi = \frac{R}{N\sigma}. \tag{3.12}$$

3.2.3 Target Material

In this experiment, a gold (Au-197) foil is used as a target material. The Au-197 is the stable isotope of gold and its natural abundance is 100%. When Au-197 absorbs neutrons through capture reactions, the radioactive isotope Au-198 is produced by the reaction ^{197}Au$(n, \gamma)^{198}$Au. Figure 3.8 shows the decay diagram of Au-198. Au-198 decays to Hg-198 by β^--decay with a half-life of 2.695 days. 99.0% decays to the first excited state of Hg-198, and populates to the ground state by emitting 411.8 keV γ-ray almost simultaneously. Therefore, the activity of the activated gold foil is generally determined by measuring β-rays with a gaseous detector, or by measuring γ-rays with a high-purity germanium (HPGe) semiconductor detector.

Fig. 3.8 Decay diagram of
Au-198

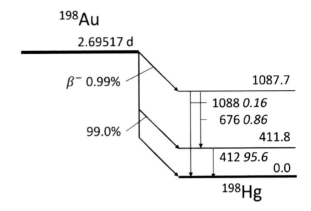

Figure 3.9 shows the cross section of the radiative capture reaction of gold. It can be seen from the figure that the cross section decreases monotonically in a straight line in the low energy region. Since the cross section is almost inversely proportional to neutron velocity v (i.e., proportional to $1/v = 1/\sqrt{E}$), this energy region is often called "$1/v$ region." In the higher energy region, there is a sharp peak (resonance peak), and the cross section changes rapidly in a narrow energy range. In the database of nuclear reaction cross sections called the Nuclear Data Library, cross section data for nuclear reactions with thermal neutrons are given for neutrons of 0.0253 eV (2200 m s^{-1}), and for Au, the activation cross section for neutrons of 0.0253 eV is 98.65 barn.

The Au is often chosen as a target material for the following reasons.

Firstly, the cross section of the radiative capture reaction is large, and the half-life of Au-198 is appropriate. As mentioned earlier, the saturated activity A_∞ is equal to the reaction rate R, and the reaction rate R is linearly proportional to the cross section

Fig. 3.9 Cross sections of
(n, γ) reactions of Au and Cd
(Data from Ref. [1])

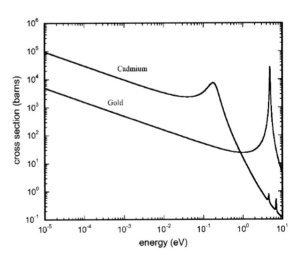

Table 3.2 Target materials used for thermal neutron flux measurements (Ref. [2])

Element	Isotope and its abundance (%)	Induced activity	Radiative capture cross section (barn)[a]	Half-life
Manganese (Mn)	^{55}Mn: 100	^{56}Mn	13.3 ± 0.2	2.5789 h
Cobalt (Co)	^{59}Co: 100	^{60}Co	37.18 ± 0.06	5.2713 y
Copper (Cu)	^{63}Cu: 69.17 ^{65}Cu: 30.83	^{64}Cu ^{66}Cu	4.50 ± 0.02 2.169 ± 0.03	12.700 h 5.120 m
Silver (Ag)	^{107}Ag: 51.839 ^{109}Ag: 48.161	^{108}Ag ^{110m}Ag	37.6 ± 1.2 4.7 ± 0.2	2.37 m 249.950 d
Indium (In)	^{113}In: 4.29 ^{115}In: 95.71	^{114m}In ^{114}In ^{116m}In ^{116}In	8.1 ± 0.8 3.9 ± 0.4 81 ± 8 40 ± 2	49.51 d 71.9 s 54.12 m 14.10 s
Dysprosium (Dy)	^{164}Dy: 28.18	^{165m}Dy ^{165}Dy	1610 ± 240 1040 ± 140	1.257 m 2.334 h
Gold (Au)	^{197}Au: 100	^{198}Au	98.65 ± 0.09	2.69517 d

[a] Cross sections are for 0.0253 eV neutrons

of the reaction. So, if the cross section is small, a large activity cannot be obtained, and sufficient counts therefore cannot be obtained in the measurement. In addition, if the half-life is too short, the activity decreases within a short time after the irradiation and that leads difficulties in a measurement. Conversely, if the half-life is too long, that also leads difficulties, because a long irradiation time is necessary to obtain a sufficient activity.

Secondly, the only natural isotope of Au is Au-197, which makes it easy to obtain a highly pure material. The high purity is advantageous because a highly pure material does not contain other isotopes or impurities that can be activated by neutrons. If non-target radionuclides are produced in the activation, their activity will interfere with the measurement. Another advantage is that thin Au foils or wires are commercially available and easy to obtain. A large neutron detector deteriorates the spatial resolution of the measurement and also disturbs the neutron flux to be measured by the detector. On the other hand, Au foils or wires can be fabricated much smaller than ordinary neutron detectors and can be used to perform measurements with good spatial resolution and without significantly affecting the neutron flux. Target materials used in the activation method are shown in Table 3.2.

3.2.4 Measurement of Thermal Neutron Flux

In the previous explanation, irradiated neutrons were assumed to be monenergistic for simplicity, but in actual irradiation environments, neutron energy is not monenergistic except in special cases, and the activation cross section varies with neutron energy as described in Fig. 3.9.

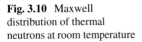

Fig. 3.10 Maxwell distribution of thermal neutrons at room temperature

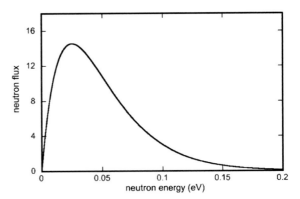

In the case of a thermal neutron reactor such as UTR-KINKI, neutron spectrum (distribution of neutron flux with respect to neutron energy) in the irradiation field where samples and detectors are installed is expressed as a superposition of the spectrum of thermal neutrons in thermal equilibrium with surrounding materials and the spectrum of epi-thermal neutrons with higher energy. The thermal neutron flux distribution $\phi(E)$ is approximated by the Maxwell distribution as shown in the following equation:

$$\phi(E) = \phi_{\text{th}} \frac{E}{(kT)^2} e^{-\frac{E}{kT}}, \tag{3.13}$$

where ϕ_{th} is the thermal neutron flux (cm^{-2} s^{-1}), k is the Boltzmann's constant (8.6173×10^{-5} eV K^{-1}), and T is the neutron temperature (K). The neutron temperature is approximately same as the moderator temperature in a system with little neutron absorption. Figure 3.10 shows the Maxwell distribution.

Taking into account the energy dependence of the activation cross section and neutron flux $\phi(E)$, the reaction rate R in Eq. (3.4) can be rewritten as follows:

$$R = N \int_0^\infty \sigma(E)\phi(E)dE. \tag{3.14}$$

When the target material is Au, we can take advantage of the fact that the activation cross section has the $1/v$ behavior in the thermal neutron region. If the cross section at an arbitrary speed v_0 (energy E_0) is σ_0, the energy-dependent cross section $\sigma(v)$ and $\sigma(E)$ can be expressed by the following equations, respectively:

$$\sigma(v) = \sigma_0 \frac{v_0}{v}, \tag{3.15}$$

$$\sigma(E) = \sigma_0 \sqrt{\frac{E_0}{E}}. \tag{3.16}$$

Equation (3.14) is then rewritten as follows:

$$R = N\bar{\sigma}\phi_{th},$$ (3.17)

where $\bar{\sigma}$ is the average activation cross section for thermal neutrons with the Maxwell distribution, and ϕ_{th} is the thermal neutron flux, and they are, respectively, given by:

$$\bar{\sigma} = \frac{\int_0^\infty \sigma(E)\phi(E)dE}{\int_0^\infty \phi(E)dE} = \frac{\sqrt{\pi}}{2}\sigma_0,$$ (3.18)

$$\phi_{th} = \int_0^\infty \phi(E)dE.$$ (3.19)

Therefore, the thermal neutron flux can be deduced from Eq. (3.17) as follows:

$$\phi_{th} = \frac{R}{N\bar{\sigma}} = \frac{2}{\sqrt{\pi}} \cdot \frac{R}{N\sigma_0}.$$ (3.20)

3.2.5 Cadmium Filter Method

As described in Sect. 3.2.4, the neutron spectrum in the irradiation field of a thermal neutron reactor is a superposition of the spectrum of thermal and epi-thermal neutrons. Therefore, the Au foil installed in the neutron irradiation field will be irradiated not only by thermal neutrons but also by epi-thermal neutrons. In particular, since the activation cross section of Au has a large resonance peak at about 4.9 eV (see Fig. 3.9), epi-thermal neutrons also activate the Au foil. Thus, the activity generated in the Au foil consists of components for thermal and epi-thermal neutrons and is given by

$$A = A_{th} + A_{ep},$$ (3.21)

where A is the total activity generated in the foil, and A_{th} and A_{ep} are components for thermal and epi-thermal neutrons, respectively.

To obtain the thermal neutron flux from the activity generated in the Au foil, it is necessary to remove the component caused by epi-thermal neutrons (A_{ep}) from the total activity (A) and extract the component for thermal neutrons (A_{th}). For such separation of activity, a method known as the "cadmium filter method" is generally used. Cadmium (Cd) is known as a strong neutron absorber, and the Cd absorption cross section is very convenient for the purpose of separating thermal and epi-thermal neutrons. As shown in Fig. 3.9, the Cd absorption cross section has a high peak for neutrons with the energy about 0.4 eV. Here, if the Au foil is covered with two Cd plates of appropriate thickness, the Cd plates act as a filter to block neutrons with the energy lower than about 0.4 eV, and the Au foil is only activated by neutrons with the energy more than about 0.4 eV. The energy of 0.4 eV is considered as the boundary where the Cd plate blocks neutrons or not. If we regard the energy as the boundary between the energy regions of thermal and epi-thermal neutrons, we can obtain the activity produced by thermal neutrons, by taking the difference between the activities produced in the Au foils covered with and without the Cd plates. This is the principle of the cadmium filter method.

In the actual irradiation experiment, two Au foils are prepared, one of them is covered with Cd plates, and the other is irradiated with neutrons without Cd plates. The difference in the activity per unit mass should be calculated for taking into account that the number of target nuclei in the two Au foils is different (the weight of the two Au foils is not equal). When the saturated activity per unit mass of the Au foil covered with Cd plates is a_c (Bq g^{-1}), and that of the bare Au foil not covered with Cd plates is a_b (Bq g^{-1}), the saturated activity per unit mass produced by thermal neutrons a_{th} (Bq g^{-1}) is given by

$$a_{th} = a_b - a_c. \tag{3.22}$$

The ratio of the activities produced in the two Au foils covered with and without Cd plates is called the "cadmium ratio." The cadmium ratio R_{Cd} can be expressed as follows:

$$R_{Cd} = \frac{a_b}{a_c}. \tag{3.23}$$

When the cadmium ratio is known, the activity of the bare gold foil a_b can solely be used to obtain the activity generated by thermal neutrons a_{th} as follows:

$$a_{th} = a_b \left(1 - \frac{1}{R_{Cd}} \right). \tag{3.24}$$

The cadmium ratio is used as an index to show the ratio of the density of thermal and epi-thermal neutrons at the irradiation position. Noteworthy is that the larger the ratio, the larger the proportion of thermal neutrons.

3.2.6 Experiment

3.2.6.1 Sample Preparation and Irradiation

(1) Prepare Au foils to be irradiated and weigh them with an electronic balance. Two Au foils are necessary for each irradiation position. Number each foil and record its weight and irradiation position.

(2) One of the two Au foils at each irradiation position is sandwiched between two Cd plates, and the other is left bare.

(3) Open the top shield closures of the reactor and place the foils at the irradiation positions in the graphite reflector of the reactor core. The positions will be appointed by a supervisor. After the foil installation is completed, the top shield closures are closed. The crane is operated by a staff member to open and close the top shield closures, so do not enter the reactor room during the crane operation.

(4) Start up the reactor, increase the power, make the reactor a critical state of 1 W in automatic operation mode, and start irradiation. The time when the reactor reaches a critical state of 1 W is defined to be the start of irradiation ($t = 0$), so the stopwatch is started at the same time as the reactor reaches a critical state of 1 W. The irradiation time (t_0) will be instructed by the supervisor.

(5) Shut down the reactor at the end of irradiation ($t = t_0$). Since the radiation dose in the reactor core immediately after shutdown is high, wait until the counting rate of the fission counter is less than 10 cps and the "Low Count Rate" indicator light turns on (usually it takes about 10 min). When the indicator light turns on, open the top shield closures again and take out the irradiated foils. Since the foils are activated (radioactive), handle them carefully by following the instructions of the supervisor.

3.2.6.2 Activity Measurement (β-Ray Measurement with a GM Counter)

(1) Apply high voltage to a GM counter. The voltage to be applied will be instructed by the supervisor on the day of the experiment.

(2) Count background radiation to obtain background counting rate.

(3) Determine the detection efficiency of the counter by counting the β-rays from a standard β-source of known activity.

(4) Since the Au-198 has a long half-life of about 2.7 days, it is not necessary to pay much attention to the decay of radioactivity after removal from the reactor. In addition, be sure to record the start time ($t = t_1$) and end time ($t = t_2$) of the measurement for each sample.

3.2.6.3 Activity Measurement (γ-Ray Measurement with an HPGe Detector)

(1) Confirm that the dewar vessel of the HPGe detector is filled with liquid nitrogen and that the detector is sufficiently cooled (it is usually prepared by staff prior to the experiment but must be checked).

(2) Apply high voltage to the HPGe detector. The applied voltage will be instructed by the supervisor on the day of the experiment.

(3) Determine the detection efficiency (peak efficiency) of the measurement system for the γ-rays emitted from Au-198 (411.8 keV) (see the supplementary explanation).

(4) Measure γ-rays from each sample and acquire a gamma spectrum with a multi-channel analyzer (MCA). Obtain the net counting rate of the photoelectric peak (it is recommended to use the function of the software that controls the MCA). Be sure to record the start time ($t = t_1$) and end time ($t = t_2$) of the measurement for each sample.

3.2.6.4   Data Analysis and Discussion

(1) Summarize the information on the standard source used for determining the detection efficiency (nuclide, source number, activity, calibration date, etc.) and obtained data such as measurement time, counts, background, etc.
(2) Summarize the foil number, weight, measured value, start time of irradiation and irradiation duration, start time and end time of the measurements, and other necessary information on each sample as a list.
(3) Calculate the saturated activity per weight (Bq g^{-1}) (reaction rate per weight) of each sample using Eq. (3.11).
(4) Calculate the activity produced by thermal neutrons and the cadmium ratio for each irradiation position using Eqs. (3.22) and (3.23), respectively.
(5) Calculate the thermal neutron flux at each irradiation position using Eq. (3.20).
(6) In the case that the measurement was performed at several irradiation positions, summarize the thermal neutron flux and the cadmium ratio at each position and plot their spatial distributions. The spatial distribution of the thermal neutron flux can be approximated by a cosine distribution when the measurement is performed at several irradiation positions in the vertical direction of the reactor.
(7) What corrections would be needed for a more precise measurement of thermal neutron flux using the activation method?

[Column] How to determine photoelectric peak efficiency

If the net count rate of the photoelectric peak measured with an HPGe detector is n (cps) and the γ emission rate of the source is I (s^{-1}), the peak efficiency ε_p is expressed by the following equation:

$$\varepsilon_p = \frac{n}{I}. \qquad (3.25)$$

The peak efficiency of an HPGe detector is complicatedly related to the geometric arrangement of the detector and γ emitter, and the energy of the γ-rays, etc. Therefore, it is generally difficult to obtain the exact peak efficiency easily.[1] In this experiment, it is necessary to determine the peak efficiency for the 411.8 keV γ-rays

---

[1] While a standard source can be regarded as a point source, a gold foil is a plane source, and it is difficult to make the geometric conditions of the detector the same as those of a standard source. By separating the source and detector, the geometrical conditions of the detector for the gold foil become closer to those for point sources, but the counting rate decreases and the statistical accuracy becomes poor.

emitted from Au-198. Since a standard source emitting 411.8 keV γ-rays is usually not available, the peak efficiency is estimated as follows:

The peak efficiencies (ε_{p1}, ε_{p2}, ...) are obtained for γ-rays of several energies (E_1, E_2, ...) using several standard γ-ray sources specified by the supervisor. Then, the results are plotted on a double logarithmic graph with the energy (denoted E) on the horizontal axis and the peak efficiency (denoted ε_p) on the vertical axis, and the first-order fitting is performed as follows:

$$\left(\log \varepsilon_p\right) = a + b(\log E), \tag{3.26}$$

where a and b are constants determined by fitting. From Eq. (3.26), we obtain the peak efficiency of the detector for the 411.8 keV γ-rays.

It is known that, in a limited range from a few hundred keV to a few MeV, the simple empirical expression in Eq. (3.26) can be used to express the relationship between the γ-ray energy and the peak efficiency.

[Column] Sum peak method

The stable isotope of Na-23 (natural abundance: 100%) absorbs thermal neutrons well and becomes the radioactive Na-24 by the (n, γ) reaction. Sodium carbonate (Na_2CO_3) powder, which is a compound containing sodium, can be encapsulated and used as a target material for the activation method.

Na-24 decays to Mg-24 by β^- emission with a half-life of 14.96 h, and two γ-rays (2.754 MeV and 1.369 MeV) are emitted in a cascade pattern immediately after the decay. When they are measured with a high-purity germanium (HPGe) semiconductor detector, the pulse height distribution shows not only the photoelectric peaks at 2.754 MeV and 1.369 MeV but also a peak at the position corresponding to the sum of the energies of the two γ-rays (4.123 MeV). The peak is called the "sum peak," which appears when the two γ-rays are detected by photoelectric interactions simultaneously. Figure 3.11 shows a typical pulse height spectrum of γ-rays from Na-24 measured with an HPGe semiconductor detector.

Fig. 3.11 Pulse height spectrum of γ-rays from Na-24 measured with HPGe detector

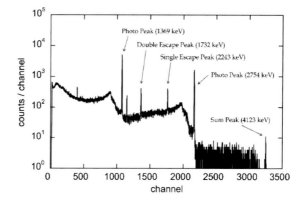

Let the Na-24 activity of a sample A_0 (Bq), and the two γ-rays emitted by the decay be γ_1 and γ_2. If the total detection efficiencies for γ_1 and γ_2 are ε_1 and ε_2, respectively, the probability that γ_1 and γ_2 are then detected simultaneously will be $\varepsilon_1\varepsilon_2$. Also, the probability that only γ_1 is detected is $(\varepsilon_1 - \varepsilon_1\varepsilon_2)$, and the probability that only γ_2 is detected is $(\varepsilon_2 - \varepsilon_1\varepsilon_2)$. Furthermore, if the probabilities of photoelectric interaction for γ_1 and γ_2 are ε_{1p} and ε_{2p}, respectively, the peak counting rate of γ_1, γ_2 and the sum peak are as follows:

$$n_1 = A_0(\varepsilon_1 - \varepsilon_1\varepsilon_2)\varepsilon_{1p} = A_0(1 - \varepsilon_2)\varepsilon_1\varepsilon_{1p}, \tag{3.27}$$

$$n_2 = A_0(\varepsilon_2 - \varepsilon_1\varepsilon_2)\varepsilon_{2p} = A_0(1 - \varepsilon_1)\varepsilon_2\varepsilon_{2p}, \tag{3.28}$$

$$n_s = A_0\varepsilon_1\varepsilon_2\varepsilon_{1p}\varepsilon_{2p}, \tag{3.29}$$

where n_1, n_2 and n_s are the peak counting rates for γ_1, γ_2 and sum peak, respectively. The total counting rate n_T is then expressed as follows:

$$n_T = A_0(\varepsilon_1 + \varepsilon_2 - \varepsilon_1\varepsilon_2). \tag{3.30}$$

All four detection efficiencies ε_1, ε_2, ε_{1p} and ε_{2p} can be eliminated from Eqs. (3.27) to (3.30), and finally the following relation is obtained:

$$A_0 = n_T + \frac{n_1 n_2}{n_s}. \tag{3.31}$$

Therefore, the Na-24 activity in the sample can be measured absolutely from the counting rate of the photoelectric peak obtained by the measurement and the total counting rate.

It is common, however, that the low pulse height part including noise in the spectrum is removed by setting the threshold level in the measurement system. In addition, since the background count must also be removed, an ingenuity is necessary to obtain the total count rate n_T.

3.3 Measurements of Neutron and γ-Ray Dose Rates

Air dose rates are measured at the measurement points around the reactor at the full power of 1 W in the UTR-KINKI operation.

Fig. 3.12 Neutron energy
spectrum ratio at the center
of the core

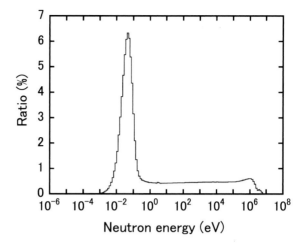

3.3.1 Production of Neutron and γ-Ray

During the reactor operation, neutrons and γ-rays are produced. The origin of
neutrons has come from fission reactions in the reactor, including prompt and delayed
neutrons. The origins of γ-rays are as follows: the generation of fission products in
the reactor; the γ-rays as a result of inelastic scattering of neutron with materials;
the γ-rays from neutron-induced radioactive materials.

The energy distribution of neutrons at the center of the core is shown in Fig. 3.12.
The energy ranges from thermal neutron to a few MeV. In the core, radiation is
generated from the activation of in-core components and fission products, and γ-rays
do not reach the reactor surface due to biological shielding.

3.3.2 Measuring Instruments and Principle of Measurement

The purpose of the measurement of air dose rates around the reactor is to confirm
that the leaked dose rate from the reactor is less than the value specified in the safety
regulations (Refs. [6, 7]). It is confirmed by the measurement that both the dose due
to external exposure of workers around the reactor (Refs. [8, 9]) is sufficiently low
and that there is no damage to the biological shielding of the reactor and no point
with a peculiarly high dose rate.

Air dose rates in the experiment will be measured using survey meters, which are
portable measuring instruments among the ambient dose equivalent rate measuring
instruments. The most commonly used survey meters for γ-rays are an ionization
chamber survey meter and a NaI(Tl) scintillation survey meter. The latter is capable
of measuring even low dose rate levels of natural background dose rate, and the range
of dose rate that can be measured is generally limited to 30 μSv h^{-1}. Since the dose

Fig. 3.13 Neutron dose
conversion coefficients per
fluence

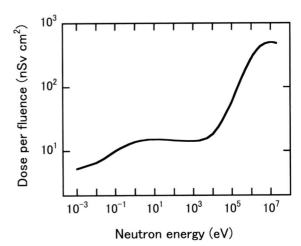

rate on the outer surface of the top shield of the reactor operating at the full power
of 1 W can exceed 30 μSv h^{-1}. In the experiment, an ionization chamber survey
meter is going to be used. Some neutron ambient dose equivalent rate meters are also
called REM counters. The radiation weighting factor used to evaluate the effective
dose is determined by a function of energy in the case of neutron exposure. Therefore,
the coefficients for converting fluence to dose differ depending on the energy of the
incident neutron. Figure 3.13 shows the coefficients for converting fluence to dose
for neutron energy.

An ideal neutron ambient dose equivalent rate detector has the energy response that
matches the curve shown in Fig. 3.13. Most of the neutron ambient dose equivalent
rate detectors are proportional counters filled with helium-3 (He-3) gas. The reaction
cross section of He-3 is shown in Fig. 3.14. As shown in Fig. 3.14, the He-3 (n,
p) reaction used for detection has a large cross section in the lower energy region.
To fit the detector response to the curve in Fig. 3.13, the detector is surrounded by
polyethylene of about 10 cm thickness. Polyethylene is composed of carbon and
hydrogen, and the fast neutron incident on the detector is slowed down and shifted to
the lower energy side by elastic scattering with hydrogen. The slowing down of fast
neutrons makes it highly sensitive to fast neutron. Meanwhile, the pulse generated by
the incident γ-rays on the detector does not contribute to the dose rate measurement
because of its low pulse height.

3.3.3   Measurement

Air dose rates will be measured at the measurement points around the reactor that is
at the full power of 1 W in UTR-KINKI. The measurement points (MP) are shown

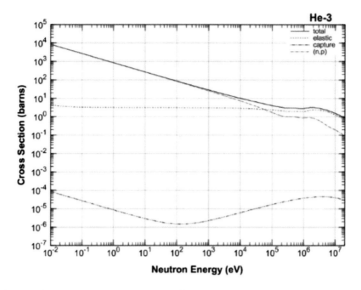

Fig. 3.14 Reaction cross section of He-3 (Data from Ref. [1])

in Fig. 3.15. The measured values will be recorded in Fig. 3.16. The location and the height of the points are as follows:

MP1: Center of the reactor top, on the surface
MP2: Off-center of the reactor top, on the surface
MP3: Top of the stairs, 0.8 m above the surface of the reactor top
MP4 to MP7: Surface of the biological shielding tank,
              0.8 m above the bottom level of the tank
MP8: Surface of the storage box, 0.8 m above the floor
MP9: Entrance of the reactor room, 0.8 m above the floor.

Fig. 3.15 Measurement points around the reactor

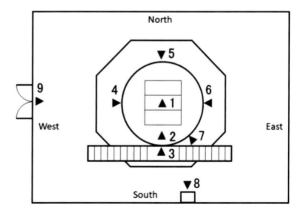

Measurement point	Dose rate (γ-ray)	Dose rate (neutron)	Sum
1			
2			
3			
4			
5			
6			
7			
8			
9			

Fig. 3.16 Data sheet of measurement results (unit: μSv h^{-1})

3.3.3.1 Measurement Preparation

(1) Ionization chamber survey meter: start up by selecting the switch from OFF to ON. Perform a zero check and set the measurement range to 10 μSv h^{-1}.
(2) Survey meter for neutron: press the power button to turn on the power.

3.3.3.2 Measurement and Recording

(1) The order of measurement is MP4 to MP9 as the first half, and the next is MP1 to MP3.
(2) Align the detection part of the survey meter with the measurement position and wait for the dose rate to settle to a certain value before reading and recording. Each reading shall be recorded with two significant digits. At measurement points MP1 to MP3, the measuring range of the ionization chamber survey meter shall be switched to the appropriate range.
(3) At each measurement point, add up γ-ray dose rates and neutron dose rates, and record the total value. In this case, the second decimal place shall be rounded up to the nearest whole number in μSv h^{-1}, and the first decimal place shall be recorded. No rounding or truncation shall be performed to ensure that the values are on the safe side.

3.3.4 Discussion

3.3.4.1 Difference Between Dose Rates at Measuring Points

Consider the reasons for (1) and (2) as follows:

(1) Why are north–south (measurement points MP5 and MP7) and east–west (measurement points MP4 and MP6) different?

(2) For measurement point MP8, why is the neutron dose rate larger than the γ-ray dose rate?

3.3.4.2   External Exposure Doses for Workers

At the full power of 1 W of UTR-KINKI, the entry to the reactor top is restricted, and then, there is no restriction of entry in the reactor room except for the reactor top. The "Regulation on Prevention of Ionizing Radiation Hazards" related to occupational safety and health stipulates that the radiation dose at the place where people always enter should be kept lower than 1 mSv per week. In the safety regulations for the reactor facility, the control target (constraint value) for the reactor side dose rate is set at 20 μSv h^{-1}. Confirm that the air dose rates at the measurement points (MP4 to MP9) do not exceed these control target values. Furthermore, discuss the reason why the control target value was set at 20 μSv h^{-1} for a 40-h work week.

3.3.4.3   Quantitative Consideration of Shielding

The distances from the center of the reactor core to each measurement point, the shielding materials and their thicknesses between them are shown in Table 3.3. Here, the material of filled sand with water is considered equivalent to concrete. Using the radiation transport calculation code PHITS (ver. 3.02) [10], we obtained numerically the dose rate (Sv h^{-1}) of incident neutron on graphite or concrete with a thickness of 150 cm. The incident neutron energy is as follows: 1 MeV, 2 MeV and 3 MeV. The shielding effects by thickness are shown in Figs. 3.17, 3.18 and 3.19, in relative values with the dose rate at the incident point as 1. The dose rates of the γ-rays generated by inelastic neutron scattering are shown by dashed lines.

Refer to Table 3.3 and Figs. 3.17, 3.18 and 3.19, and estimate the effective energy of the neutron at the center of the reactor core to be closest to 1 MeV, 2 MeV or

Table 3.3 Material and thickness of shielding and distance from the center of the core

Measurement point	Material and thickness of shielding	Distance from the center of the reactor core
1	Graphite (122 cm) Concrete (45 cm)	197 cm
4	Graphite (51 cm) (Filled sand + water) (142 cm)	198 cm
5	Graphite (51 cm) (Filled sand + water) (127 cm)	198 cm
6	Graphite (51 cm) (Filled sand + water) (142 cm)	198 cm
7	Graphite (51 cm) (Filled sand + water) (127 cm)	198 cm

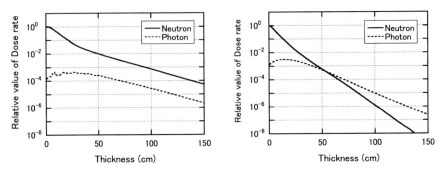

Fig. 3.17 Shielding against 1 MeV neutron radiation (left: graphite; right: concrete)

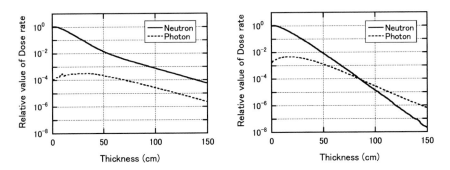

Fig. 3.18 Shielding against 2 MeV neutron radiation (left: graphite; right: concrete)

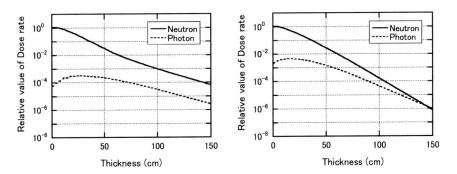

Fig. 3.19 Shielding against 3 MeV neutron radiation (left: graphite; right: concrete)

3 MeV, based on the ratio of the air dose rate of the neutron to the γ-ray at each measurement point.

[Column] Leakage γ-ray spectrometry

As described in Sect. 3.3.1, γ-rays are emitted from in-core activated components and fission products, which rarely reach the surface of the reactor because γ-rays are shielded by the biological shield. Moreover, γ-rays are emitted by the reactions where low-energy neutrons are captured by stable nuclei, and as a result, the excited nuclei immediately de-excite. The γ-rays emitted in the process are called the prompt γ-rays. Since the prompt γ-rays have a relatively high energy and are penetrating, they constitute a major component of the γ-rays leaking from the biological shield. High-purity germanium detectors (HPGe detector) have excellent energy resolution and are suitable for measuring the energy spectrum of the prompt γ-rays.

(1) HPGe detector

There are two types of HPGe detectors; the one is a p-type detector and the other is an n-type detector. The p-type is generally used for the measurement of γ-rays with the energy more than 40 keV. The p-type coaxial HPGe detector has a $p +$ layer formed by boron ion implantation in the central hole and an $n +$ layer formed by lithium diffusion on the surface. A depletion layer is formed by applying reverse bias between $p +$ and $n +$ layers and is sensitive to γ-rays. The detection efficiency depends on the size of the depletion layer, and a detector with a thick depletion layer is desirable to obtain sufficient detection efficiency for high-energy γ-rays.

The measurement of γ-rays with the HPGe detector is based on the interaction between γ-rays and Ge crystal. When secondary electrons are generated by the interaction, an electron–hole pair is generated in the depletion layer, which can be read out as an electrical signal. The greatest characteristic of the detector is that γ-ray energy can be determined with high accuracy. This is called that the detector has good energy resolution. The energy resolution is expressed as the full width at half maximum (FWHM; Fig. 3.20), which is defined as the width at the P/2 of the peak with the height P. Since γ-ray energy can be directly read out as the number of electron–hole pairs, and the average energy to generate an electron–hole pair is as small as 2.9 eV, there is little fluctuation in the number of electron–hole pair production, and good energy resolution can be realized. The FWHM of the HPGe detector is usually smaller than 2 keV for the 1332.5 keV γ-rays emitted from Co-60.

Since the band gap of the HPGe detector is small, the detector needs to be cooled to an operating temperature by liquid nitrogen or an electric cooler in order to reduce the effect of thermal noise caused by electrons exceeding the band gap.

Fig. 3.20 Full width at half maximum of a peak

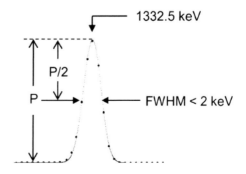

(2) Measurement system

As shown in Fig. 3.21, when radiation enters a detector and gives its energy, the energy is read out as an electrical signal. Figure 3.21 shows the waveform of the signal after signal processing (preamplifier and main amplifier) is conducted. Radiation can be counted one by one, and one signal (pulse) is generated when radiation is detected. The pulse height V is proportional to the energy E of the secondary electron generated by the interaction between γ-rays and Ge crystal. The energy spectrum shown in Fig. 3.22 is obtained by discriminating this signal with a pulse height analyzer (multichannel analyzer: MCA).

(3) Leakage γ-ray spectrometry

When measuring γ-ray spectrum leaking from the biological shield of the reactor, the detector is placed as close as possible to the surface of the shield tank. After an energy calibration, the preset time of the MCA is set and the measurement is

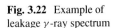
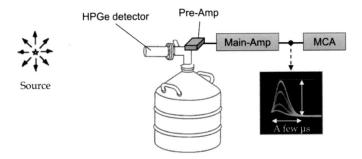
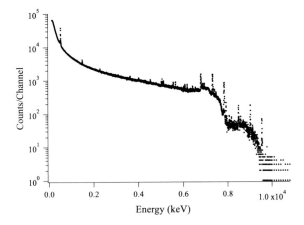

Fig. 3.21 Configuration of γ-ray spectrum measurement system

Fig. 3.22 Example of leakage γ-ray spectrum

performed for a certain period of time. Figure 3.22 shows an example of the leakage γ-ray spectrum. As shown in Fig. 3.22, the energy of the leakage γ-rays is relatively high and reaches 10 MeV. Therefore, it is necessary to perform the energy calibration by reducing the gain of the main amplifier sufficiently so that a wide energy spectrum can be obtained.

(4) Data analysis

Calculate the channels of the peaks that appear in the spectrum using the functions of the software provided with MCA. From the relationship between the MCA channel and the energy obtained by the energy calibration, convert the channels of the peak to the energy and list them. From the results of energy distribution, find out what kind of capture reaction caused the identified peak.

When identifying the peak of the γ-spectrum, it is recommended to use the database of prompt γ-rays shown as follows:

IAEA, "Database for Prompt Gamma-ray Neutron Activation Analysis, Most Intense Gamma Rays."

https://www-nds.iaea.org/pgaa/Annex1/Table-IV-Chap7.pdf, Accessed 1 July 2022.

3.4 Neutron Radiography

3.4.1 Introduction

Radiography is the technology that uses the penetrating effect of radiation to see through the interior of an object. The most familiar radiography technology is the X-ray radiography used in medical examinations. As the non-destructive inspection technique that does not damage or destroy the object, the X-ray technology is used not only for medical purposes but also in a wide variety of fields, including the inspection of industrial products, the investigation of cultural properties, science and engineering research, and security checks at the airport.

A medical X-ray image is a picture of the inside of a human body taken with X-rays, and such an image can be taken with not only X-rays but also neutron beams. The technique of imaging a subject using neutron beams is called neutron radiography or neutron imaging.

When the same subject is photographed using X-rays and neutron beams, the resulting images are very different. This is due to the fact that X-rays and neutrons have different materials to interact with. When X-ray radiography and neutron radiography are used together for the same subject, complementary information about the components and structures inside the object can be obtained. Furthermore, different techniques are required to make an image with X-rays and neutrons.

The purpose of the neutron radiography experiment is to understand the principle of neutron radiography and the imaging technique by taking a picture of a subject using neutrons generated in the reactor. The same subject will be also imaged with X-rays, and the two images obtained by X-rays and neutrons will be compared to see how the differences in the interaction of the two radiations with materials appear in the images.

3.4.2 Attenuation of X-rays and Neutrons in Matter

X-rays are absorbed or scattered and attenuated in matter by three kinds of interactions (photoelectric effect, Compton scattering and pair production). Among these, the photoelectric effect and Compton scattering are interactions with electrons in materials, and it can be imagined that atoms with more electrons, i.e., atoms with larger atomic numbers attenuate X-rays more in the energy region where these interactions dominate. In fact, as shown in Fig. 3.23, the mass attenuation coefficient of X-rays ($cm^2 g^{-1}$) increases monotonically with atomic number.

When a subject is irradiated with X-rays, X-rays are shielded by absorption and scattering in materials with large atomic numbers, whereas X-rays are transmitted through materials with small atomic numbers. Thus, the shadows of materials with large atomic number contained in the subject appears in the image as if it were a shadow picture. The reason why hard tissues, such as bones and teeth inside a human body, appear clearly on radiographs used in medical examinations is that hard tissues contain calcium (Ca) and have a relatively higher atomic number than soft tissues such as muscle and fat (the effective atomic number of hard tissues is about twice that of soft tissues).

On the other hand, unlike X-rays, neutrons are attenuated by direct interaction (absorption or scattering) with nuclei. The mass attenuation coefficient of thermal neutrons does not change monotonically with atomic number as in the case of X-rays, because the probability of the interaction varies with each nuclide. Figure 3.23

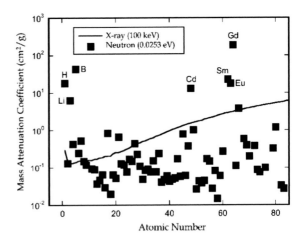

Fig. 3.23 Mass attenuation coefficient for X-rays and thermal neutrons (Refs. [1, 11])

also shows the mass attenuation coefficients of thermal neutrons against atomic number. The mass attenuation coefficient of thermal neutrons varies greatly with atomic number and is large even for elements with small atomic numbers, such as hydrogen (H), lithium (Li) and boron (B). There are also other elements, such as cadmium (Cd) and gadolinium (Gd), that have particularly large mass attenuation coefficient. In addition, neutrons can easily penetrate heavy metals such as iron and lead, which are difficult for X-rays to penetrate. Therefore, neutron radiography can be used to visualize and inspect hydrogen-rich materials, including water, oil and plastics in structures made of heavy metals.

3.4.3   Imaging Plate

Photographic film has been used for X-ray diagnosis for a long time because the film is an integral two-dimensional X-ray detector which can produce images easily using established techniques. In 1981, the Fujifilm Corporation developed the imaging plate (IP; Ref. [12]) as a new computed radiography technology to replace photographic X-ray films. Since IPs are approximately 100 times more sensitive than conventional photographic films for X-rays, the radiation exposure in X-ray diagnosis was greatly reduced by using IPs. In addition, there is no need to develop a film to obtain an image as is the case with conventional photographic film, and the images can be stored for a long time without deterioration because they are acquired as digital data. The IP has various excellent characteristics as a two-dimensional radiation detector, such as the ability to perform quantitative analysis by image processing and the large dynamic range (the IP has a dynamic range of about five orders of magnitude, whereas the conventional photographic film has about two orders of magnitude).

The principle of X-ray imaging with IP is very different from that of photo-graphic film and is based on a phenomenon called photostimulated luminescence (PSL). When a special material called a stimulable phosphor is irradiated with radi-ations, the material absorbs and accumulates the energy of radiation. Then, when the phosphor is irradiated with a laser beam, the phosphor emits the stored energy as the photostimulated luminescence. Figure 3.24 shows a photograph and the structure of IP, which is composed of a thin plastic sheet coated with BaFBr:Eu and covered with a protective film. After the IP is irradiated with X-rays, the surface of the IP is scanned with a He–Ne laser (633 nm), and the luminescence (UV light) is detected and converted to electric signal with a photomultiplier tube. The signals are then digitized and processed with a computer to make a two-dimensional image. The resolution of the image can be increased to several tens of micrometers. In addition, by irradiating the IP with visible light for about ten minutes after the readout, all the information stored in the IP is erased and the IP can be used again.

Since the IP for X-rays does not contain any neutron-sensitive components, it cannot be used for neutron radiography as it is. Therefore, neutron-sensitive IPs have been developed by mixing Gd as a neutron converter in the stimulable phosphor. The Gd has a very large (n, γ) reaction cross section for thermal neutrons, and

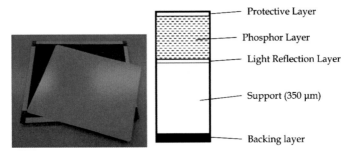

Fig. 3.24   IP (left) and cross-sectional structure (right)

absorbs thermal neutrons well and emits prompt γ-rays and conversion electrons. The conversion electrons in particular can give energy to the vicinity of the Gd crystal that absorbed neutrons and the energy is stored in the IP as a neutron image, which can be used for neutron radiography.

3.4.4   Neutron Radiography Facility

There are several neutron sources used for neutron radiography, including research reactors, accelerator neutron sources and radioisotopes (RIs). Among these, a research reactor is a stable source with high-intensity neutrons and has been used widely for neutron radiography. Accelerator neutron sources have also been widely used in recent years because their performance has been improved. Here, we describe the neutron radiography facility of UTR-KINKI (Ref. [13]), which is used for the experiment.

The upper part of the reactor is closed by concrete top shield closures during the reactor operation. The top shield closures can be replaced with upper irradiation facilities depending on the purpose of the experiment. There are several types of upper irradiation facilities, one of which is the B-facility for neutron radiography. The other facilities are the A-facility for small animal irradiation and the C-facility for the insertion of experimental samples to the reactor core during the reactor operation. Figure 3.25 shows a photograph of the neutron radiography facility (B-facility) in UTR-KINKI.

Figure 3.26 shows a cross-sectional view of the neutron radiography facility in UTR-KINKI. The collimator (neutron guide tube) in the neutron radiography facility is located just above the central stringer hole, and the neutron beam is derived from the area where thermal neutron flux is maximum. The collimator is made of mortar with lithium fluoride, which absorbs thermal neutrons efficiently, and the neutrons that collide with the inner wall of the collimator are absorbed so that thermal neutron beams are at the exit of the collimator as parallel as possible.

Fig. 3.25 Neutron radiography facility in UTR-KINKI (© AERI, Kindai University. All rights reserved)

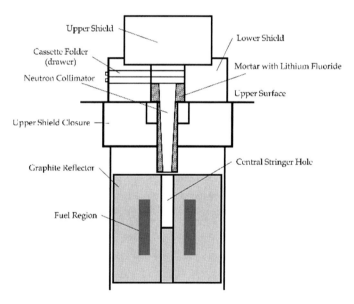

Fig. 3.26 Cross-sectional view of neutron radiography facility in UTR-KINKI

The performance of a collimator can be expressed by the collimator ratio (L/D), where L is the length of the collimator and D is the diameter of the collimator inlet. In the case of UTR-KINKI, $L = 92$ cm and $D = 9$ cm, so the collimator ratio is 10.2. In general, the collimator ratio of a neutron radiography facility should be more than

10, and preferably more than 100, so the collimator ratio of UTR-KINKI is close to the lower limit.

The lower shield is installed on the collimator, and the upper shield is installed on the lower shield. The lower shield has a cassette folder in which the subject and neutron IP are installed, and can be put in and out like a drawer during reactor operation. Since the diameter of the collimator exit is 20 cm, the diameter of the irradiation field is about 20 cm. In addition, the thermal neutron flux in the irradiation field at a critical state of 1 W is about 1×10^4 cm^{-2} s^{-1}.

Since the full power of UTR-KINKI is 1 W, and the thermal neutron flux is smaller than that of other neutron radiography facilities, a long irradiation time is required to obtain the neutron fluence required for imaging. In the case of neutron IPs, the typical irradiation time to obtain a still image is about 15 min.

3.4.5 X-ray Generator

X-rays used for common X-ray diagnosis are generated by an X-ray tube. Figure 3.27 shows the configuration of an X-ray generator using an X-ray tube. The voltage applied between the cathode and anode is called the tube voltage, and a high voltage of several tens to hundreds of kV is applied. The current carried between the cathode and anode is called the tube current. When thermo electrons generated from a filament (cathode) heated to a high temperature are accelerated by the tube voltage and collide with a metal target (anode), X-rays are generated by bremsstrahlung.

The efficiency for the conversion of electrical power to X-rays is small, and most of the kinetic energy of electrons is converted to heat when they collide with a metal target, and only a small amount is converted to X-ray energy. The metal target is

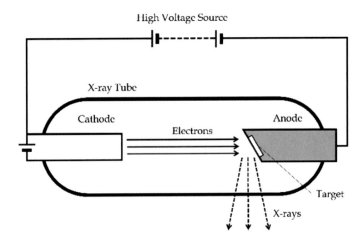

Fig. 3.27 Configuration of X-ray generator

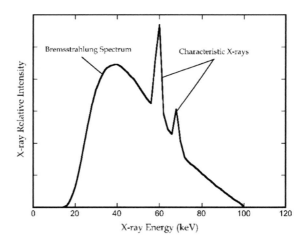

Fig. 3.28 Energy spectrum of X-rays generated by X-ray tube (tube voltage: 100 kV, target: W, absorber: Al 1.0 mm)

then heated during X-ray generation, so it is often cooled by water. In addition, since X-rays are generated by bremsstrahlung, it is more efficient to use a material with a large atomic number as a target to generate X-rays. Thus, a metal with a large atomic number and a high melting point is advantageous for the metal target, and tungsten (W) is often used.

The energy spectrum of X-rays emitted from an X-ray tube consists of a continuous spectrum of bremsstrahlung and a peak of characteristic X-rays emitted from a metal target. Figure 3.28 shows the energy spectrum of X-rays emitted from an X-ray tube. The maximum energy of the continuous spectrum is determined by the tube voltage of the X-ray tube, i.e., the acceleration voltage of the electrons.

The intensity of the generated X-rays is determined by the current flowing between the electrodes of the X-ray tube, i.e., the tube current, because the intensity increases with the number of electrons striking the metal target per unit time. However, since the efficiency of X-ray generation by bremsstrahlung increases with the energy of the electrons, the X-ray intensity is higher with a higher tube voltage for the same tube current.

The image contrast of radiographs is related to the energy of X-rays. The higher the energy of X-rays, the higher the transmission power, and thus the lower the image contrast. In addition, radiographic density is determined by X-ray intensity and X-ray irradiation time. The product of the tube current (mA) and the irradiation time (s), mAs (milliampere-seconds), is used as the measure of the amount of X-rays produced by X-ray tube. Moreover, with the same amount of mAs, an image with the same radiographic density can be obtained even if the combination of the tube current and irradiation time is different.

3.4.6   *Experiment*

(1) Prepare several subjects to be radiographed and place them in a 20 cm diameter circle drawn in the center of an aluminium (Al) plate. Once the subjects are placed, fix them with scotch tape.

(2) Insert X-ray IPs and neutron IPs in an eraser for about 10 min to erase any remaining image information. Put each IP into a cassette, which is a light-shielded container for an IP. It is important not to confuse the IPs for X-rays and neutrons.

(3) Put the Al plate with subjects on the cassette containing the X-ray IP, and irradiate X-rays with an X-ray generator. Set the tube voltage at 40 kV and the tube current at 1 mA, and irradiate X-rays for 3 s.

(4) Operate the reactor with the neutron radiography facility in automatic operation mode at a critical state of 1 W. Place the cassette containing the neutron IP and the Al plate with subjects in the cassette folder and insert it into the lower shield. The irradiation time is 15 min.

(5) After the irradiation is completed, set the IPs in a reader and read out the image.

3.4.7   *Discussion*

Figure 3.29 shows examples of radiographic image. Compare two images radiographed with X-rays and neutrons, and discuss how the difference in the interaction between X-rays/neutrons and materials is shown in the images.

Fig. 3.29   Radiographic images of same objects using neutrons (left) and X-rays (right)

References

1. Shibata K, Iwamoto O, Nakagawa T et al (2011) JENDL-4.0: a new library for nuclear science and engineering. J Nucl Sci Technol 48:1–30
2. The Japan Radioisotope Association (2011) Radioisotope pocket data book 11th edn. Tokyo, Japan (in Japanese)
3. Naito K (1978) Genshiro kagaku (Nuclear reactor chemistry), nuclear engineering series 4. University of Tokyo Press, Tokyo, Japan (in Japanese)
4. ATOMICA The principle of silicon semiconductor production by neutron irradiation. (in Japanese). https://atomica.jaea.go.jp/data/detail/dat_detail_08-04-01-25.html. Accessed 1 July 2022
5. The Japan Radioisotope Association (2020) Mippusengen no kiso (Fundamentals of sealed sources), 7th edn. Maruzen Publishing, Tokyo, Japan (in Japanese)
6. Guide for joint use of the Kindai University Reactor. (in Japanese). https://www.kindai.ac.jp/files/rd/research-center/aeri/guide/external-use/outside4.pdf. Accessed 1 July 2022
7. Atomic Energy Research Institute, Kindai University, Safety regulations for the nuclear reactor facility. (in Japanese)
8. Nuclear Regulation Authority of Japan (2000) Notice on quantity of radioisotopes, quantities etc. of radiation-emitting isotopes, public notice of the science and technology agency, no. 5, Oct. 23, 2000. https://www.nsr.go.jp/data/000182250.pdf. Accessed 1 July 2022
9. Shibata T ed (2021) Hoshasen gairon (Introduction to radiation) 13th edn. Tsusho-sangyo-kenkyu-sha, Tokyo, Japan (in Japanese)
10. Sato T, Iwamoto Y, Hashimoto H et al (2018) Features of particle and heavy ion transport code system (PHITS) version 3.02. J Nucl Sci Technol 55:684–690
11. Hubbell JH, Seltzer SM (1995) Tables of X-ray mass attenuation coefficients and mass energy-absorption coefficients from 1 keV to 20 MeV for elements Z = 1 to 92 and 48, additional substances of dosimetric interest. NIST standard reference database 126 https://www.nist.gov/pml/x-ray-mass-attenuation-coefficients. Accessed 1 July 2022
12. Fujifilm Corporation, Fujifilm development of a new X-ray diagnostic imaging system "FCR." (in Japanese). https://www.fujifilm.co.jp/corporate/aboutus/history/ayumi/dai5-08.html. Accessed 1 July 2022
13. Niwa T, Koga T, Morishima Y et al (1987) Characteristics of neutron radiography facility constructed at Kinki University reactor. J Atomic Energy Soc Japan 29:904–912 (in Japanese)

Chapter 4
Test and Research Reactors in Japan

Abstract Test and research reactors are nuclear reactors that are designed to conduct education and research using radiation, such as neutrons, differing from light-water reactors that are designed to generate electricity. Test and research reactors are generally called "research reactors" including small zero-power reactors called critical assemblies. Detailed information on test and research reactors, such as their structure, purpose, characteristics, classification and utilization, is described in this chapter, using the general term of "research reactors."

Keyword Test and research reactors in Japan

4.1 Overview

Research reactors are designed to produce neutrons suitable for science and engineering research, material irradiation, radioisotope (RI) production and medical irradiation. Light-water reactors (LWRs) produce steam at high temperature and high pressure converted into heat energy to generate electricity, and conversely, the heat energy generated in research reactors is not used to generate electricity. Since research reactors were originally designed to use neutrons, the volume of fuel region in research reactors is relatively small, comparing with that of LWRs with a diameter of about 3.5 m and a height of about 4 m. So why is the fuel region in research reactors small? In general, the smaller the size of the core, the easier neutrons generated in the core leak out of the core. In research reactors, the fuel region is designed to be small in size and high in power density so that the neutrons generated in the core can leak out of the core, to use as many neutrons as possible at experimental holes and ports.

Here, let's find out "how many research reactors are there in Japan?" In Japan, there are a total of 22 research reactors, including 13 decommissioned reactors. As of July 2022, only six research reactors are in operation.

Several research reactors owned by private companies and universities have been operated since the 1960s. The only research reactor owned by a private company is now the Toshiba Nuclear Critical Assembly (NCA), and three research reactors owned by universities are follows: the University Teaching and Research Reactor of

G. Wakabayashi et al., *Introduction to Nuclear Reactor Experiments*,
https://doi.org/10.1007/978-981-19-6589-0_4

Kindai University (UTR-KINKI); the Kyoto University research Reactor (KUR) and the Kyoto University Critical Assembly (KUCA) of Kyoto University. Considering the cost of compliance with new regulations, maintenance and operation after the accident at the Fukushima Daiichi Nuclear Power Plant, it is becoming more and more difficult for private companies and universities to own their own research reactors.

The Japan Atomic Energy Agency (JAEA) has many research reactors, including the Japan Research Reactor-3 (JRR-3) that mainly utilizes neutrons, the Nuclear Safety Research Reactor (NSRR) and the Static Experiment Critical Assembly (STACY) for safety researches in LWRs, and the High Temperature engineering Test Reactor (HTTR) and "Joyo" for new reactors including High Temperature Gas-cooled Reactors (HTGRs) and Fast Reactors (FRs), respectively. In the research reactors owned by Toshiba, Kindai University and Kyoto University, human resource development programs for education and training purposes are actively conducted, except for KUR.

4.2   Characteristics

In Sect. 4.1, it was explained that research reactors are designed to leak neutrons easily to the outside, and the core of the research reactor is small and has a high-power density to utilize as many neutrons as possible. Although research reactors use fuels with higher uranium enrichment (U-235 enrichment: < 20 wt%) than power reactors (U-235 enrichment: < 5 wt%), a thermal power of research reactors is lower (mostly < 100 MW) than that of LWRs (300,000 to 1,300,000 kW). In the past, research reactors used fuels with U-235 enrichment up to about 90 wt% (highly enriched), and highly enriched fuels are being converted into lower enrichment fuels (U-235 enrichment less than 20 wt%) except for some research reactors.

The nuclear fuel of many research reactors is uranium-silicide-aluminum (U_3Si_2 -Al) or uranium-aluminide-aluminum (UAlx-Al) alloy plate fuel. The alloy plate fuel is called MTR (Materials Testing Reactor)-type fuel. The coolant of the MTR is low temperature and low pressure, and the aluminum alloy is often used for the core components because of its low neutron absorption. Most of the beamline facilities and material irradiation facilities are available for public use, and the operation of research reactors is usually cyclic (11 cycles: 1 to 4 weeks).

Research reactors are designed to have higher neutron flux (about 2.0×10^{14} s^{-1} cm^{-2}) than LWRs, lower thermal power (1 W to 20 MW; LWRs have a maximum of about 4000 MW), and lower reactor coolant temperature and pressure. Here, the Al and other materials with low neutron absorption are mainly used for the structural materials of the core. In addition, there are many irradiation holes and experimental holes for neutron beam extraction. The fuels, moderators and coolants used in research reactors are summarized in Table 4.1, and the purpose of the utilization is summarized in Table 4.2.

On the other hand, LWRs generate high-temperature and high-pressure steam using the thermal energy that is available to rotate a turbine and generate electricity.

Table 4.1 Fuels, moderators and coolants of research reactors in Japan

	Thermal power	Fuel	Moderator	Coolant	Purpose
JRR-3	20 MW	U_3Si_2-Al	H_2O	H_2O	Beam use
NSRR	300 kW	U-ZrH	H_2O	H_2O	Accident studies
HTTR	30 MW	UO_2	Graphite	He gas	Development of HTGRs
"Joyo"	140 MW	U–Pu	–	Metal Na	Development of FRs
UTR	1 W	U-Al	H_2O	–	Personnel development
KUR	5 MW	U_3Si_2-Al	H_2O	H_2O	Multipurpose
KUCA	100 W	U-Al	H_2O (Polyethylene)	–	Research on reactor physics

Table 4.2 Summary of utilization in research reactors

Field of utilization	Contents	Example
Research on reactor physics	Basic research on nuclear reactors and nuclear materials	– Research on core configuration and fuel arrangement of new type reactor – Research on volume reduction and low toxicity of radioactive materials
Beam utilization (Material irradiation)	Analysis of durability of core structural materials to radiation	Verification of the safety of nuclear reactors and satellite components
Beam utilization (Activation analysis)	Elemental analysis of substances	– Analysis of impurities contained in semiconductors – Analysis of hazardous substances etc. contained in agricultural products
Beam utilization (Neutron radiography)	Neutron transmission photography and animation	Visualization of the inside of an automobile engine (Check engine oil behavior)
Beam utilization (Neutron scattering and diffraction analysis)	Analysis of the arrangement structure of atoms and molecules	Development of polymer materials such as polymers and surfactants
Medical irradiation	Boron neutron capture therapy	Cancer treatment
Production of RI	Production of radioisotopes	– Diagnosis and treatment of cancer (molybdenum, etc.) – Non-destructive testing of reactor components (Iridium)
Personnel development	Experimental education using a research reactor	Education and training of undergraduate and graduate students, and junior faculty

LWRs are also large in size and have high thermal power because they are designed to give priority to economic efficiency. Low-enriched uranium dioxide (UO$_2$) (U-235 enrichment: 2 to 5 wt%) is used as nuclear fuel. Zirconium alloys (in fuel assemblies) and austenitic stainless steels (in-core tanks, piping, etc.) are often used as core components because of their low neutron absorption and high corrosion resistance in high-temperature water.

Comparing research reactors (Table 4.1) and LWRs, both have a similarity in the use of light water as moderators and coolants, and the enrichment of the fuel, U-235, is different. Noteworthy is that the core temperature of LWRs is in the range between 250 and 300 °C, and the core temperature of research reactors depends on the purpose of utilization. Incidentally, since UTR-KINKI and KUCA are operated at a thermal power lower than 1 W, as shown in Table 4.1, the increase in core temperature is almost negligible, and the cores are always operated at room temperature. The maximum temperature of the KUR that is operated at the full power of 5 MW is about 40 °C. Table 4.3 shows a comparison between UTR-KINKI and LWR (Kansai Electric Power Co., Inc.: KEPCO, Ohi units 3 and 4).

Table 4.3 Comparison between UTR-KINKI and power reactor

		UTR-KINKI	KEPCO power station, Ohi units 3 and 4 (PWR)
Power	Thermal power Electric power	1 W –	Approx. 3.4 million kW Approx. 1.1 million kW
In-core environment	Pressure temperature	Atmospheric pressure Normal temperature	Approx. 157 atm (coolant pressure) Approx. 307 °C (coolant temperature)
Reactor core	Number of fuel assemblies Core size	12 1.1 × 1.4 × 1.2 m	193 Approx. 3.4 m diameter × approx. 3.7 m height
Shutdown	Control rod	Cadmium (four plates)	Silver, Indium, Cadmium (53 rods)
Cooling	Coolant	–	Cooling water flow 81 s^{-1} m^3
Cladding	Fuel coating material Reactor vessel Containment vessel	Aluminum – –	Zircaloy4 Approx. 4.4 m diameter & 13.0 m height Approx. 4.3 m diameter & approx. 6.6 m height

4.3 Classification

4.3.1 Classification by Utilization Purpose

As indices to classify research reactors, neutron spectrum, thermal power, nuclear fuels, moderators, coolants and the purposes of utilization were introduced in Sect. 4.2, and neutron characteristics that are closely related to the primary purpose of utilization are also described here.

The main purposes of utilization of research reactors can be divided into the following categories:

- Education and training
- Basic research
- Material irradiation
- Production and development research of RI
- Medical irradiation
- Engineering research.

Based on the purposes of utilization, research reactors are classified under the following common names:

- "Experimental beam reactor" for using neutrons to conduct researches in physics, chemistry, biology and medicine
- "Material testing reactor" for reactor fuel and structural material irradiation
- "Reactor for RI production"
- "Reactors for education and training"
- "Engineering test reactors," "experimental reactors" and "prototype reactors" for the development of next-generation reactors.

Many research reactors are constructed for multiple purposes, including beam experiment, RI production and material irradiation. Education and training reactors are mainly installed in universities and have a wide range of thermal power from 1 W to 5 MW to support various reactor experiments. The reactor physics experiments in critical assemblies are omitted, and please see Ref. [1].

4.3.2 Classification by Design

(1) Swimming pool-type research reactor

The swimming pool-type research reactor (Refs. [2, 3]) is a research reactor in which the core is installed at the bottom of the pool water (light water) used as neutron moderators, reactor coolants, reactor reflectors and radiation shielding materials. The core can be seen directly through the water, and the area around the core is wide enough for easy operation. The Japan Research Reactor-4 (JRR-4) is a typical reactor of the type. The MTR-type fuel is used in the reactor, and U_3Si_2-Al (silicide) and UAlx-Al (aluminide) alloy plate fuel elements are standard. JRR-3 M, JRR-4 and NSRR (TRIGA-ACPR type; 300 kW, up to 23,000 MW in pulsed operation; Ref. [4]) are classified as swimming pool open-type research reactors.

(2) Tank-type research reactor

In high-power reactors, such as the Japan Material Test Reactor (JMTR) (Ref. [5]), the core is installed in a tank (reactor vessel) to suppress the boiling of reactor coolant, called a swimming pool tank type or simply a tank type (Ref. [3]). Light water, heavy water, beryllium (Be), Al and graphite are used as reactor reflectors, and the Be and Al are used in JMTR.

4.4 Applications

4.4.1 Education and Training

The purposes of education and training reactors are to contribute to the operation and management of nuclear reactors, to acquire skills for radiation control and to promote understanding of principles and characteristics of nuclear reactors. Education and training reactors have also been used to train reactor professionals, including researchers, engineers and workers, and to promote understanding of nuclear reactors among the general public, including government officials, media and students involved in the nuclear field. The Rikkyo University reactor, which had operated for 40 years since 1961, was decommissioned in December 2011. The Musashi Institute of Technology (currently Tokyo City University) reactor was the TRIGA-II research reactor that had reached a first critical state in 1963 was decommissioned in 2003. Furthermore, the operation of the University of Tokyo reactor "Yayoi" has also been suspended. Now, only UTR-KINKI and KUCA can be used for the purpose of education and training. The Nuclear Human Resource Development Center of JAEA has conducted reactor operation training, control rod calibration experiments and various characteristic measurements, with the use of JAEA's research reactors, such as JRR-4. The JRR-4 has opened its doors to domestic and foreign countries for education and training. The JAEA has announced, however, the plan to decommission JRR-4 as a part of business rationalization.

In the 1960s, major heavy electric power manufacturers had their own research reactors to establish reactor design technology and to train nuclear engineers. The Hitachi Training Reactor (HTR) that was the first reactor in the private sector had been constructed by Hitachi, Ltd., and used for development and research of reactors, activation analysis, radioisotope production, material irradiation and medical irradiation. Also, all decommissioning work of HTR was completed in October 2005.

Now (as of July 2022), the NCA owned by Toshiba is being prepared for decommissioning, while NCA is the only research reactor (criticality experimental device) owned by a private company.

4.4.2 Beam Experiments

Neutrons generated in the reactor core are guided through a horizontal experimental hole setting in the reflector (mainly heavy water) region of the reactor and taken out of the reactor for utilization. The experimental beam reactors in Japan are as follows: KUR, NSRR and JRR-3.

The KUR was shut down for about two years from March 2006 and converted from highly-enriched uranium fuel into low-enriched uranium fuel. Because of neutrons together with the low energy γ-ray, the KUR has a heavy-water thermal neutron facility suitable for medical–biological irradiation, a cold neutron source facility and a long-wavelength neutron facility suitable for molecular structure studies of biological materials and polymeric materials, a low-temperature irradiation facility and a quasi-monoenergetic neutron experimental facility. Neutron scattering experiments, structural analyses by the neutron radiography (NRG) and trace element analyses by prompt γ-ray analysis have been carried out, and activation analyses and fission track dating have been carried out using the irradiation facilities.

The NSRR has been operated for more than 30 years since June 1975, and 3154 pulse operations and 1350 fuel irradiation experiments have been performed. Through the results of these experiments, the NRG is expected to contribute to the understanding of the in-core condition of the Fukushima Daiichi Nuclear Power Plant and to achieve the decommissioning at an early stage. The NRG has also been performed in UTR-KINKI, and excellent images have been obtained, although the neutron dose is weak.

In beam application researches using JRR-3, various neutron scattering instruments have been used to elucidate the structure of living organisms and materials in the fields of life science and material science. Residual stress measurements are also carried out for industrial applications. In addition, NRG has been conducted for researches of nuclear power plants and nuclear reactor safety. In the field of prompt γ-ray analyses, meteorites, archaeological samples and biological samples are analyzed.

4.4.3 Material Irradiation

Research reactors that have several irradiation facilities are called irradiation test reactors, including capsule irradiation facilities, hydraulic rabbit irradiation facilities and boiling water capsule irradiation facilities for irradiating nuclear fuels for irradiating reactor component materials and fusion reactor materials, under various conditions. Several irradiation experiments are then carried out by using vertical irradiation holes equipped with the core and reflector regions of the reactor.

Many irradiation experiments are needed to make noteworthy progress in the development and research of LWR technologies, including the countermeasure against the aging of current LWRs, the performance enhancement and safety evaluation of UO_2 and MOX fuels (Mixed Oxide: mixture of uranium and plutonium fuels). Of research reactors, JMTR is designed to generate a larger amount of neutrons than LWRs so that irradiation tests can be carried out quickly (accelerated irradiation tests). In addition, many tests have been performed in JMTR as follows: irradiation tests for high burnup of LWR fuels, irradiation tests to investigate the irradiation behavior of structural materials as a countermeasure for the aging of LWRs, irradiation tests for the development of new reactors and coupling irradiation tests between the experimental fast reactor "Joyo" and the JMTR. The JMTR operation was stopped in August 2006, and the renewal of the reactor and the improvement of the irradiation facilities have been started. As for the future studies, JMTR has a large number of demands for irradiation of LWR fuels and materials, production of RIs for medical and industrial use and irradiation of large-diameter neutron transmutation doping (NTD) used for hybrid vehicles. The decommissioning of JMTR is, however, being prepared in JAEA (as of July 2022).

In JRR-3, accelerated irradiation tests of 100 GWd·t^{-1} and irradiation tests of Rock-like plutonium fuels with excellent nuclear nonproliferation and environmental safety have been carried out to study the high-burnup region of LWR fuels. Moreover, irradiation tests of HTTR and fusion reactor materials have been continued in the material irradiation tests.

4.4.4   Isotope Production

Of isotopes possessed by an element, the isotopes that emit some kind of radiation are called RIs, when the atomic nucleus decays, because the atomic nucleus is unstable. The time that the decay is half is then called the half-life. The radiation emitted by the decay is used for various purposes, such as medical diagnosis and cancer treatment. For example, in terms of technetium-99m (Tc-99m), the isotope of Tc-99m is used for imaging tests to depict organs, including bones, kidneys, lungs, thyroid gland, liver and spleen by using the substance that specifically binds to a specific receptor as a label, taking advantage of the characteristics of emitting only γ-ray without emitting β-ray.

Almost all of the RIs widely used in the fields of medicine, industry, agriculture, forestry and fisheries are produced in research reactors or cyclotron accelerators. The types of RIs produced by research reactors and accelerators have a wide range of variations, because the energy range of nuclear reactions induced by the RI production varies in accordance with the types of RIs. Typical RIs produced in research reactors are Tc-99m, iridium-192 (Ir-192) and gold-198 (Au-198) for medical diagnosis and treatment, Ir-192 and cobalt-60 (Co-60) for non-destructive testing and thickness gauges, and tritium and carbon-14 (C-14) for labeled compounds. As for the Ir-192 and Au-198 that used for cancer treatment, although research reactors (JRR-3 and JMTR) had provided 100% of the domestic demand, the supply of the two RIs has been stagnant because the reactors have been shut down since the Great East Japan Earthquake in 2011.

Meanwhile, it is impossible to supply all the RIs required for domestic use, and many RIs are imported. A substantial part of the imported RIs is radiopharmaceuticals. Among the RIs, the Tc-99m that emits low-energy γ-ray with a short half-life is administered to the human body as a radiopharmaceutical, and is used for diagnosis and functional testing of many disease sites, including bone, brain and liver cancer, covering about 80% of RIs used in medicine.

4.4.5 Activation Analysis

When a sample is irradiated with neutrons in a nuclear reactor, radionuclides are produced in the sample by neutron capture reactions. By measuring the type of nuclide made and the amount of nuclide produced, the type and concentration of the elements contained in the sample can be determined. The measurement method is called "activation analysis." The activation analysis is also used for irradiation by charged particles accelerated by a cyclotron or bremsstrahlung radiation by electrons accelerated from an electron linear accelerator. Here, the activation analysis is the elemental analysis method using nuclear reactions. Moreover, the activation analysis is applicable to the analysis of samples of various compositions and has the characteristics of non-destructive multielement simultaneous determination and high-sensitivity analysis. Another part of the characteristics is that the activation analysis is less affected by the matrix than other instrumental analysis methods and has a wide range of concentrations that can be determined.

The activation analysis method, performed by irradiating neutrons from a nuclear reactor, has been highly evaluated as a high-precision trace element analysis method. For example, the method is applied to as follows: the analysis of heavy metal elements, the investigation of environmental pollution caused by oil spills, the determination of trace elements contained in environmental indicator marine products (seaweed, squid, scallops, plankton, etc.) and the determination of trace elements contained in soil, hot spring water, submarine sediments and meteorites.

4.4.6 Medical Irradiation

The boron neutron capture therapy (BNCT) has been performed in the medical field at the heavy-water thermal neutron beam facility at KUR. In BNCT, boron is delivered to the tumor site through blood via an intravenous drip, and neutrons from a nuclear reactor are applied to the affected area. Then, the boron is transmuted into lithium, and α-ray is simultaneously emitted. At KUR, thermal and epi-thermal neutrons are used for the treatment of brain tumors, and the treatment techniques that maximizes the effects of BNCT by thermal and epi-thermal neutrons have been developed. Moreover, BNCT is particularly effective for malignant tumors with indistinct borders that cannot be treated by surgery or conventional radiation therapy. Japan has played a leading role in the development of BNCT and has irradiated brain tumors at JRR-4, and skin cancer, lung cancer, liver cancer and other organ cancers in addition to brain tumors at KUR.

4.4.7 Silicon Doping

An example of producing materials for semiconductors by neutron irradiation is the production of silicon (Si) semiconductor devices by the NTD method (or silicon doping) (Ref. [6]). The Si-30 that exists in about 3% of Si is transformed into Si-31 (half-life: 2.62 h) through absorption reactions with neutrons, and transmuted into phosphorus-31 (P-31: a stable isotope) through β-ray decays. Since the P-31 is uniformly dispersed in Si and the concentration of P-31 can be precisely controlled, a high-quality semiconductor material is finally produced by the transmutation of Si-30 into P-31. The Si semiconductors irradiated in nuclear reactors show better electrical properties than those produced by the gas diffusion method. Therefore, a large-diameter (150 mm) semiconductor Si is made and is considered necessary in the semiconductor industry for use in large ultra-high-voltage thyristors.

4.4.8 Others

(1) High-temperature heat utilization reactor

The first criticality of the HTTR was achieved in November 1998, and the effective full power of 30 MW and reactor outlet coolant temperature of 850 °C were achieved in December 2001. The HTTR was constructed to establish and upgrade the basis of HTGR technology. In addition to the accumulation of operational data, control rod withdrawal tests have been conducted to demonstrate the safety characteristics of HTGRs, simulating reactivity-induced events and flow reduction tests simulating coolant flow reduction. Moreover, in 2004, the iodine (I) and sulfur (S) compound process (IS process) that is a thermochemical water-splitting process succeeded in producing hydrogen continuously at $0.03 \text{ m}^3 \cdot \text{h}^{-1}$ for one week. Furthermore, the design concept of power reactors with a small size, a high temperature and a direct gas turbine has been developed by utilizing the results of the HTTR design, construction, operation and safety demonstration tests (Ref. [7]).

(2) Reactors for the development of new-type nuclear reactors

The development of fast reactors and new-type conversion reactors has been conducted in Japan, considering the development of next-generation reactors. The first stage of fast reactor development was involved in irradiation tests at the experimental reactor "Joyo," the second stage was in operation tests at the prototype reactor "Monju," and the third stage was in global tests at the demonstration reactor. The research and development of new-type conversion reactors started from the construction of prototype reactor "Fugen" without the experimental reactor stage.

The experimental fast reactor "Joyo" is a sodium-cooled fast neutron reactor, which has been designed, constructed and operated with domestic technology as a national project. The "Joyo" has been operated for a quarter of a century with a breeder core (MK-I core) and an irradiation core (MK-II core), increasing technical experiences on fast breeder reactors. In 2000, the modification of the MK-III core was started, and the modification work was completed in November 2003, and the full-scale operation of MK-III started in May 2004.

The "Monju" is the only fast breeder reactor capable of generating electricity in Japan. On December 8, 1995, the accident was, however, occurred that sodium reflectors leaked from the outlet piping of the intermediate heat exchanger of the secondary cooling system. Moreover, in August 2010, "Monju" was shut down again due to a fall of an in-core relay device for fuel exchange. After that, a decommissioning of the plant was decided due to the discovery of inspection leaks in many components.

The new-type conversion reactor "Fugen" is a heavy-water-moderated, boiling, light-water-cooled (pressure tube type) reactor, starting operation in March 1979 and operating smoothly for 24 years until the end of the operation in March 2003. The "Fugen" was designed, constructed and operated by domestic technology. The "Fugen" was the first power reactor operated in the world, using MOX fuel in full scale, confirming the reliability of the pressure tube, establishing the operation method, and upgrading the operation and maintenance technologies.

References

1. Atomic Energy Society of Japan, Reactor Physics Division ed (2019) Physics of reactors (textbook of reactor physics: elementary edition). Atomic Energy Society of Japan, Tokyo, Japan. (in Japanese) https://rpg.jaea.go.jp/else/rpd/others/study/text_each.html. Accessed 1 July 2022
2. ATOMICA Pool reactor. (in Japanese) https://atomica.jaea.go.jp/dic/detail/dic_detail_1141.html. Accessed 1 July 2022
3. ATOMICA Outline of the research reactor (03-04-01-01). (in Japanese) https://atomica.jaea.go.jp/data/detail/dat_detail_03-04-01-01.html. Accessed 1 July 2022
4. ATOMICA Nuclear safety research reactor (NSRR) (03-04-02-5). (in Japanese) https://atomica.jaea.go.jp/data/detail/dat_detail_03-04-02-05.html. Accessed 1 July 2022
5. ATOMICA JMTR (03-04-02-04). (in Japanese) https://atomica.jaea.go.jp/data/detail/dat_detail_03-04-02-04.html. Accessed 1 July 2022

6. ATOMICA Silicon doping. (in Japanese) https://atomica.jaea.go.jp/dic/detail/dic_detail_1755. html. Accessed 1 July 2022
7. ATOMICA How to use the high temperature gas-cooled reactor (03-03-05-03). (in Japanese) https://atomica.jaea.go.jp/data/detail/dat_detail_03-03-05-03.html. Accessed 1 July 2022

Chapter 5
Nuclear Instrumentation

Abstract Instrumentation for reactor control and protection needs to cover a wide range of measurement targets. In addition to neutron flux measurements for monitoring reactor power, a variety of measurements is required, such as coolant temperature, pressure and flow rate measurements, depending on reactor power and type. These measurements can be roughly divided into two categories. One is nuclear instrumentation, which measures neutrons and γ-rays, and the other is process instrumentation, which is used for general industrial measurements such as temperature and pressure measurements. In a zero-power reactor such as UTR-KINKI, nuclear instrumentation plays a crucial role. This chapter describes characteristics and compositions of nuclear instrumentation, including radiation detectors used for UTR-KINKI as an example.

Keywords Nuclear instrumentation · Fission counter · BF-3 proportional counter · Boron-lined proportional counter · Compensated ionization chamber · Uncompensated ionization chamber · Start-up range · Intermediate range · Power range · Safety channel · Start-up channel · Intermediate channel · Power channel · Lin-N meter · Log-N meter · Period meter

5.1 Introduction

A variety of information is necessary for the safe and stable operation of a nuclear reactor. In addition to measuring neutrons in the reactor to monitor the reactor power, more information is necessary on the temperature, pressure, flow rate and water level of the coolant in a high-power reactor. Other measurements include those for detecting impurities in the coolant and fuel failure, and there are measurement targets and techniques. Although the classification of these measurements is not strictly defined, the instrumentation used in general industrial measurements that does not involve measurements of neutrons and γ-rays is called the process instrumentation (non-nuclear instrumentation), and the instrumentation specific to nuclear technology, which measures neutrons and γ-rays, is called the nuclear instrumentation.

© The Author(s) 2023
G. Wakabayashi et al., *Introduction to Nuclear Reactor Experiments*,
https://doi.org/10.1007/978-981-19-6589-0_5

For a zero-power reactor such as UTR-KINKI, nuclear instrumentation mostly satisfies requirements for the control of the reactor. In UTR-KINKI, information on reactor power is obtained by five neutron detectors installed on the reflector of the reactor. The process instrumentation installed in UTR-KINKI consists of a seismic detector, a moderator thermometer and the water level gauge of the biological shield tank. In this chapter, we focus on nuclear instrumentation and introduce neutron detectors used for nuclear instrumentation, and then explain the nuclear instrumentation of UTR-KINKI.

Power reactors have special neutron detectors installed in the reactor core (in-core detectors) in addition to out-core detectors to obtain information such as spatial distribution of neutron flux. Process instrumentation also plays an important role in power reactors, and readers who want to know about these measurements are advised to see the references at the end of the chapter.

5.2 Nuclear Instrumentation

In the operation of nuclear reactors, the highest priority is to ensure safety in all processes from start-up to shutdown of the reactor. Moreover, the reliability of the measurement system is most required for nuclear instrumentation, unlike that for general industrial measurement or experimental research. Neutron detectors for nuclear instrumentation are chosen to be simple in structure and detection principle and to have few failures, rather than highly accurate and complicated detectors. In addition, redundancy should be provided especially for detector systems that are directly related to safety, and detectors with the same function should be used in several independent channels simultaneously to prevent complete loss of function.

Another characteristic of nuclear instrumentation is that the measurement range is extremely wide compared with general radiation measurements. In general, the neutron flux in a nuclear reactor varies with about eight to ten orders of magnitude during the period from start-up to operation at a full power, and one type of neutron detector cannot cover the entire measurement range. Then, the power range of the reactor is usually divided into three ranges and the information from the detectors in charge of each range is combined to obtain continuous information over the whole power range. In addition, the measurement of each range is made to overlap each other so that there is no discontinuity in the measurement when moving from one range to the other. The power range of a reactor is generally divided as follows:

(1) Startup range

The range in which a neutron source is inserted into a reactor to start up the operation. It is also called the source range.

(2) Intermediate range

The range in which the control rods are withdrawn and the power of the reactor is gradually increased. The exponentially increasing power is monitored.

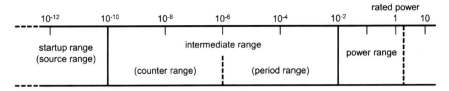

Fig. 5.1 General power range of a research reactor

(3) Power range

The range in which the reactor is operated at a full power ranging between 1 and 100%. Automatic operation is performed in this range.

The corresponding neutron flux to these power ranges depends on the full power of the reactor. Figure 5.1 shows a conceptual diagram of the power ranges. The source range and the intermediate range are sometimes referred to together as the startup range.

5.3 Neutron Detectors for Nuclear Instrumentation

The thermal power (W) of a nuclear reactor is the sum of the energy generated per unit time ($J s^{-1}$) from fission reactions in the reactor. Since fission reactions are triggered by neutrons in the reactor, the thermal power of the reactor is proportional to the number of neutrons in the reactor, which is the neutron flux as a specific physical quantity. The thermal power of the reactor can be then monitored by installing a neutron detector at a position where a sufficient number of neutrons proportional to the average neutron flux in the reactor are incident and detecting them. The obtained information on the neutron flux is used not only for reactor operation control but also for safety by producing scram signals. Since fission reactions are triggered by thermal neutrons in thermal neutron reactors, thermal neutron detectors are used for the nuclear instrumentation.

When various types of thermal neutron detectors have been developed and used for general radiation measurement applications, gaseous detectors are mainly used for nuclear instrumentation. This is because existing gaseous detectors for thermal neutron measurement are suitable for nuclear instrumentation for their simple structure, high reliability, stable operation for a long period of time, wide operating range, excellent n-γ discrimination performance, resistance to radiation damage, and low susceptibility to activation by neutrons. Research and development of scintillation and semiconductor detectors for neutron measurement continue, but they are not used for nuclear instrumentation because many scintillation detectors are sensitive to γ-rays and false signals from photomultiplier tubes cannot be ignored under a high radiation environment, and semiconductor detectors are vulnerable to radiation damage.

Since the object of measurement is a quantity proportional to the neutron flux, information on the energy of neutrons is not necessary, and information on the number of neutrons incident on the detector is sufficient.

5.3.1 BF-3 Proportional Counters

The detector that is filled with boron trifluoride (BF-3) gas as a detector gas is called the BF-3 proportional counter. Figure 5.2 shows the structure of a typical BF-3 proportional counter. Since neutrons do not have an electric charge, they cannot ionize the detector gas directly. An isotope of boron, B-10 (natural isotope abundance: 19.9%), has a large (n, α) reaction cross section for thermal neutrons (3837 barns for 0.0253 eV neutron), and the following charged-particle production reaction occurs:

$$^{10}B + n \rightarrow {}^{7}Li + {}^{4}He.$$

In this reaction, 94% of the produced lithium-7 (Li-7) nuclei are left in the excited state of 0.482 MeV, and 6% are in the ground state. By using this reaction, thermal neutrons incident on the BF-3 proportional counter react with B-10 in the gas and are converted into two charged particles, Li-7 and helium-4 (He-4) nuclei, which ionize the gas. It is the same principle as that of the ordinary proportional counter that a signal is generated in the process of collecting electrons and ions generated in the gas at the electrodes. In other words, the BF-3 gas plays two roles of converting thermal neutrons into charged particles and of acting as a detector gas of the proportional counter.

The Q-value of the $^{10}B(n, \alpha)^{7}Li$ reaction is 2.310 MeV when Li-7 is left in the excited state and 2.792 MeV when it is left in the ground state. Since the energy of thermal neutrons is almost negligible: much smaller than 2.310 and 2.792 MeV, the sum of kinetic energies of Li-7 and He-4 is 2.310 MeV or 2.792 MeV, which is given to the BF-3 gas. This means that the output signal of the BF-3 proportional counter does not reflect the incident neutron energy, but only the Q-value of the

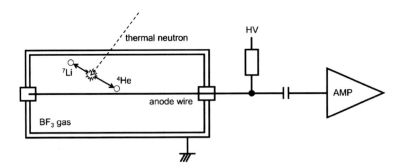

Fig. 5.2 Structure of BF$_3$ proportional counter

reaction. Since the signal originated from the Q-value is larger than that from γ-rays or electronic noise, it is easy to discriminate only the signals caused by the neutrons.

The BF-3 proportional counter is usually a tube made of a metal cylinder with a thin wire stretched in the center. A high voltage is applied between the wire (anode) and the tube wall (cathode), and pulse signals are generated by incident thermal neutrons. The sensitivity depends on the volume of the tube and the gas pressure. Since the BF-3 proportional counter is generally highly sensitive, it is mainly used in a pulse mode in a startup range.

[Example] Calculate the kinetic energies of the Li-7 and He-4 nuclei produced by the $^{10}B(n, \alpha)^7Li$ reaction, assuming that the Li-7 nucleus produced by the reaction is left in the excited state, and the Q-value of the reaction is 2.310 MeV.

Since the energy of thermal neutrons (~ 0.0253 eV) is almost negligible compared to the Q-value of the reaction, the B-10 nucleus and the thermal neutron can be considered to have been at rest in the initial state, and at some point they reacted, emitting Li and He nuclei with large kinetic energy in opposite directions. The reason why the two nuclei are emitted in opposite directions is that the sum of the momentum of thermal neutrons and the B-10 nucleus is zero due to the conservation of momentum, since they are originally at rest.

If the masses of the Li-7 and He-4 nuclei are m_{Li} and m_α, and the kinetic energies are E_{Li} and E_α, respectively, the following relations can be obtained from the conservations of energy and momentum:

$$E_{Li} + E_\alpha = Q, \tag{5.1}$$

$$\sqrt{2m_{Li}E_{Li}} = \sqrt{2m_\alpha E_\alpha}. \tag{5.2}$$

From Eq. (5.2), we get

$$\frac{E_{Li}}{E_\alpha} = \frac{m_\alpha}{m_{Li}}. \tag{5.3}$$

From Eq. (5.3), the ratio of E_{Li} to E_α is the inverse ratio of the masses of Li-7 and He-4 nuclei. As the ratio of m_{Li} to m_α can be regarded as 7/4 from their mass numbers, the ratio of E_{Li} to E_α is 4/7.

Therefore, we obtain the energies of ^7Li and ^4He nuclei as follows:

$$E_{Li} = \frac{m_\alpha}{m_{Li} + m_\alpha} Q \approx \frac{4}{7+4} \times 2.310 = 0.840 \text{ MeV}, \tag{5.4}$$

$$E_\alpha = \frac{m_{Li}}{m_{Li} + m_\alpha} Q \approx \frac{7}{7+4} \times 2.310 = 1.470 \text{ MeV}. \tag{5.5}$$

[Column] Pulse height spectrum of BF-3 proportional counter

When the output from a BF-3 proportional counter is analyzed with a multichannel analyzer (MCA), a pulse height spectrum will have a complicated structure as shown in Fig. 5.3. This characteristic structure of the pulse height spectrum can be explained by a phenomenon called the wall effect of the BF-3 proportional counter.

What would the pulse height spectrum look like if we had a very large BF-3 proportional counter? In this case, the total energy of He-4 and Li-7 produced in $^{10}B(n, \alpha)^{7}Li$ reactions would be fully absorbed in the gas, and the pulse height spectrum should show two peaks at the channels corresponding to 2.79 MeV when Li-7 is in the ground state (6%) and 2.31 MeV when it is in the excited state (94%), as shown in Fig. 5.4. The difference in the height of the peaks reflects the probability that Li-7 populates in the ground state or excited state.

The diameter of the BF-3 proportional counter is typically on the order of centimeters, and the volume of the counter is not large compared to the range of He-4 and Li-7 nuclei produced in the gas. Then, if the reaction occurs near the inner wall of the counter, either He-4 or Li-7 will collide with the wall and be absorbed (remember

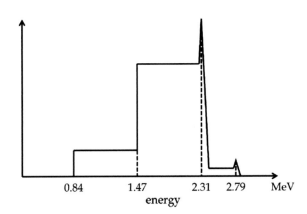

Fig. 5.3 Pulse height spectrum of BF-3 proportional counter

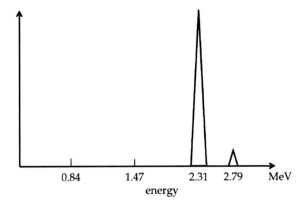

Fig. 5.4 Ideal pulse height spectrum from a large BF-3 proportional counter with 2.31 MeV (He-4) and 2.79 MeV (Li-7) nuclei fully absorbed in the gas

that He-4 and Li-7 are always emitted in opposite directions from the conservation law of momentum), and a part of the energy may not be deposited in the gas. Let us assume that Li-7 is left in the excited state, and focus on the He-4 produced in the reaction occurred near the inner wall. As shown in Fig. 5.5a, if the reaction occurs on the inner surface of the tube, Li-7 deposits all its energy in the gas while He-4 is absorbed by the wall without depositing its energy. In this case, the only energy deposited in the gas is the energy of Li-7, which means that the energy 0.84 MeV is deposited in the gas. Next, as shown in Fig. 5.5b, if the reaction occurs at a distance anywhere within the range of He-4 from the inner wall surface, all of the energy of Li-7 and part of the energy of He-4 will be absorbed in the gas. As shown in Fig. 5.5c, when the reaction occurs at a distance larger than the range of He-4 from the inner wall surface, the total energy of Li-7 and He-4, 2.31 MeV, is absorbed in the gas. Therefore, the pulse height spectrum produced by the reactions that occur at a distance within the range of He-4 from the inner wall will be a uniformly distributed rectangle spectrum from 0.84 MeV to 2.31 MeV, as shown in Fig. 5.6. As shown in Fig. 5.7, the pulse height spectrum from the reactions that occur at a distance within the range of Li-7 will also be a uniformly distributed rectangle spectrum from 1.47 MeV to 2.31 MeV.

As a result, a pulse height spectrum from a BF-3 proportional counter is a superposition of the following three spectra:

(1) The peak that is produced when all the energies of both Li-7 and He-4 are absorbed in the gas
(2) The rectangular distribution that is produced when the part of the energy of He-4 is absorbed in the inner wall
(3) The rectangular distribution that is produced when the part of the energy of Li-7 is absorbed in the inner wall.

Fig. 5.5 He-4 generated from the reaction near the inner wall of the BF-3 proportional counter absorbed by the tube wall

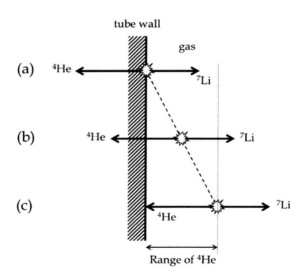

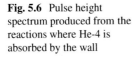

Fig. 5.6 Pulse height spectrum produced from the reactions where He-4 is absorbed by the wall

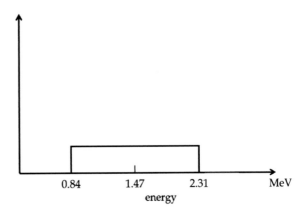

Fig. 5.7 Pulse height spectrum produced from the reactions where Li-7 is absorbed by the wall

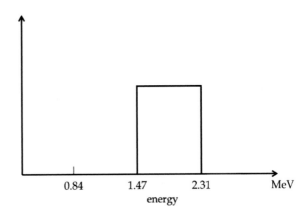

The three spectra are produced for the ground state and the excited state of Li-7, and the resulting spectrum will have the structure shown in Fig. 5.3.

5.3.2   Boron-Lined Proportional Counters

The boron-lined proportional counter is a proportional counter whose inner wall is coated with B-10-enriched boron. In the same manner as BF-3 proportional counters, the $^{10}B(n, \alpha)^7Li$ reaction by incident thermal neutrons is used to produce charged particles, ^7Li and ^4He, to ionize the gas. Unlike the BF-3 proportional counter, only one of the two charged particles can deposit its energy in the gas because the products are emitted oppositely. When one product is emitted into the gas, the other product is always absorbed in the coating layer and the wall. The efficiency of boron-lined counters can be improved as the thickness of the boron-coated layer is increased. The thickness, however, cannot be thicker than the range of the product because the products produced in the layer farther than the range from the inner surface cannot

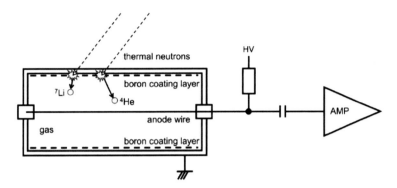

Fig. 5.8 Structure of boron-lined proportional counter

reach the gas and are never detected. Therefore, the maximum thickness of the coating will be determined by the range of ^{4}He, which is on the order of 1 mg · cm^{-2}. The advantage of boron-lined proportional counters over BF-3 proportional counters is that it can use more suitable proportional gas than BF-3 gas. Boron-lined proportional counters are also used in pulse mode in a startup range. Figure 5.8 shows the structure of a typical boron-lined proportional counter.

5.3.3 Fission Counters

The fission counter (FC) is a gaseous detector that converts thermal neutrons into charged particles using fission reactions. In other words, thermal neutrons are detected when fission fragments produced in fission reactions ionize the detector gas.

The surface of the electrodes of FC is coated with U_3O_8 enriched with U-235, and argon (Ar) is used as the detector gas. The energy generated by fission reactions of U-235 is quite large, about 200 MeV, of which about 170 MeV is the kinetic energy of fission fragments. Since the energy of fission fragments is large and the specific ionization (the number of ion pairs produced per unit path when a charged particle travels through a material) is also large, a sufficiently large charge can be obtained without gas amplification, and the detector can be operated as an ionization chamber.

As in the case of boron-lined proportional counters described in Sect. 5.3.2, since two fission fragments are also directed oppositely, when one fission fragment ionizes the gas, the other is absorbed in the coating layer. Then, only one of the two fission fragments will always impart energy to the gas. If the thickness of the coating layer is increased to gain a high efficiency, the fraction of the energy absorbed in the coating layer increases, and the average energy imparted to the gas decreases.

Since U-235 and U-238, which are isotopes of uranium, are alpha emitting radioisotopes, alpha particles are always emitted from the coating layer and causes background events. Meanwhile, the energy of fission fragments is more than ten times larger than that of alpha particles, while the energy of alpha particles is about 5 MeV. Therefore, it is easy to discriminate the pulse height based on the difference in the magnitude of the signals, and, as a result, only the signals from neutrons can be counted.

Although the efficiency of FC is generally inferior to that of BF-3 proportional counters, FC is suitable for measurements at high counting rates because the rise time of the signal is short, ranging between 0.1 and 0.3 μ s.

5.3.4   Ionization Chambers

The ionization chamber is the simplest detector in terms of detection principle and structure among various radiation detectors, and is suitable for nuclear instrumentation from the viewpoint of reliability. In ranges such as the intermediate range and the power range, where neutron flux is larger and counting rate is too high to measure in the pulse mode, the ionization chamber is used in the current mode. In the ionization chamber for neutron measurement, the B-10-enriched boron is deposited on the electrodes, and the He-4 and Li-7 nuclei emitted from the $^{10}B(n, \alpha)^7Li$ reactions ionize the gas.

Since the ionization chamber for nuclear instrumentation is used in an environment where neutrons and γ-rays are mixed, both neutrons and γ-rays produce electric charges in the chamber. When used in the current mode, neutrons and γ-rays contribute to the output current because neutrons and γ-rays cannot be discriminated by pulse height unlike in the case of pulse mode detectors.

The compensated ionization chamber (CIC) is the ionization chamber that has a mechanism to remove the contribution of γ-rays from the output current. The principle of the operation is shown in Fig. 5.9. The CIC consists of two ionization chambers sharing one electrode, and only the electrode of one of two chambers is coated with the enriched B-10. The two chambers are sensitive to γ-rays, whereas one of them is also sensitive to neutrons. By adjusting the sensitivity of two chambers to γ-rays to be the same, only the current from the neutron contribution can be extracted.

In an environment where the contribution of γ-rays is small and negligible, or where reliability is more important than accuracy, such as detectors used for a safety channel, the uncompensated ionization chamber (UIC) is then used. UIC is an ionization chamber with a simple structure whose electrodes are coated with boron to make it sensitive to neutrons.

5.4   Compositions of Nuclear Instrumentation

Neutron detectors used in nuclear instrumentation are classified into four channels according to the power range to be monitored and roles: (1) startup channel, (2) intermediate channel, (3) power channel and (4) safety channel.

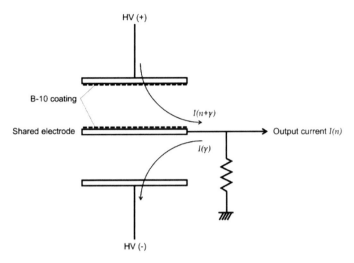

Fig. 5.9 Operation principle of compensated ionization chamber (CIC)

In this section, the power range to be monitored by the nuclear instrumentation of UTR-KINKI is introduced, as shown in Fig. 5.10, and compositions of the nuclear instrumentation are presented, as shown in Fig. 5.11. Moreover, in the following, the nuclear instrumentation is explained by using several channels used in UTR-KINKI as an example.

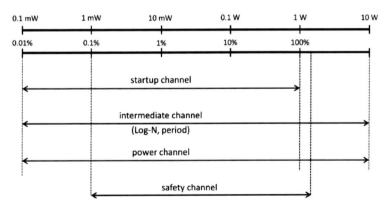

Fig. 5.10 Measurement power range of nuclear instrumentation of UTR-KINKI

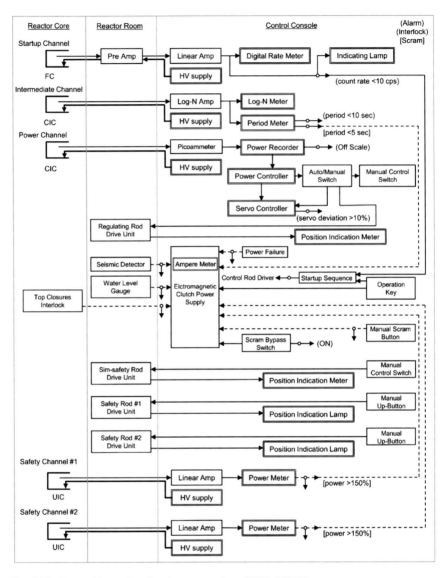

Fig. 5.11 Compositions of nuclear instrumentation of UTR-KINKI

5.4.1   Startup Channel

Startup channel is used in the startup range where a neutron source is inserted into the core to start up the reactor. The BF-3 proportional counter, the boron-lined proportional counter or FC may be used for the startup channel. Since the neutron flux is sufficiently small in the startup range, any of these detectors can be used in the pulse mode. The signals from the detectors are amplified and shaped by a preamplifier and a linear amplifier, and the signals caused by γ-rays and electronic noise are then removed by a pulse height discriminator, and, as a result, neutron signals are counted and displayed.

The FC is used as the startup channel of UTR-KINKI, and the neutron counting rate (cps) is displayed on the digital rate meter in the control console. The signals from the startup channel is also used as one of the conditions for the startup sequence of the reactor. When the counting rate is lower than 10 cps, the withdrawal of control rods is prohibited as "Low Count Rate".

5.4.2 Intermediate Channel

Intermediate channel is used to measure the power and reactor period in the process of increasing the reactor power by withdrawing the control rods, and the CIC is used as neutron detectors in the current mode. In the intermediate channel, the power, which varies exponentially over a wide range from the startup range to the power range, needs to be monitored on a single scale. For this purpose, the current signal from the CIC is converted into a voltage signal proportional to the logarithm of the input current by a logarithmic amplifier and displayed on a logarithmic power meter (Log-N meter).

The output from the logarithmic amplifier is also used for a period meter. When the power of the reactor increases exponentially, neutron flux n increases as

$$n = n_0 e^{\frac{t}{T}}, \tag{5.6}$$

where T is the reactor period and is defined as the time required to change the power by a factor of e. When the signal proportional to n is input to a logarithmic amplifier, the output is obtained by taking the logarithm of both sides in Eq. (5.6),

$$\ln n = \ln n_0 + \frac{t}{T}. \tag{5.7}$$

Differentiating Eq. (5.7) gives

$$\frac{d}{dt}(\ln n) = \frac{1}{T}, \tag{5.8}$$

and the period T is obtained from this relationship. If the output signal from the logarithmic amplifier is input to a differentiation circuit, and the output signal from the differentiation circuit is displayed on a reciprocal scale, the display indicates then a period meter.

The CIC is used for the intermediate channel of UTR-KINKI, and the output current (A) and reactor period (s) are displayed on the logarithmic power meter

(Log-N meter) and period meter of the control console, respectively. The output from the period meter is also used to trigger the signals of alarm and scram. When the period is shorter than 10 s, alarm is triggered and the buzzer sounds, and when the period is shorter than 5 s, a scram signal is generated and the reactor is shut down immediately.

5.4.3 Power Channel

Power channel is used to monitor the reactor power ranging between 1 and 100%, and is also used for the automatic control of operation to maintain the power at the demand level. The CIC or UIC is used as a neutron detector in the current mode. The output current from the chamber is amplified by a linear amplifier and converted to a current signal or a voltage signal proportional to the input current, which is generally displayed on a linear power meter (Lin-N meter).

UTR-KINKI uses the CIC as a neutron detector for the power channel, and, unlike the general configuration described above, the output current from the CIC is directly measured by a picoammeter (an ampere meter for small current measurement) and is displayed on the linear power meter. The output current is continuously measured in the entire range from the startup range to the power range to monitor the power. Since the output current varies greatly over five decades ranging between 10^{-11} and 10^{-7} A, the measurement is performed by switching the measurement range as the power increases. The relationship between the readings of the linear power meter and the power of the reactor is calibrated in advance, and the power of the reactor is 1 W when the output current is 5.72×10^{-8} A.

The signal from the linear power meter is input to the power recorder (chart recorder), which constantly records the change in the reactor power on a chart paper during the reactor operation. When the reading of the linear power meter exceeds the measurement range, and the indication of the recorder is out of the display range (over-range), an alarm is triggered as "power recorder off scale." Furthermore, the signal from the power recorder is used for automatic control of reactor operation, and is input to the servo controller for automatic operation. The servo controller is equipped with a switch to select automatic or manual operation. When the switch is selected to the automatic operation, the regulating rod (RR) is positioned by the proportional–integral–deviate (PID) control, and automatic operation is performed to maintain the reactor power at the demand level. The deviation (%) between the demand levels set by the servo controller and the actual reactor power is displayed on the servo deviation meter. Here, if the deviation exceeds 10% during the automatic operation, an alarm is triggered as "servo deviation over 10%".

5.4.4 Safety Channel

Safety channel monitors the abnormal increase of neutron flux for reactor protection. Neutron detectors for safety channel are required to have high reliability and quick response rather than accuracy. To improve the reliability, several independent detectors are used simultaneously, and the system is multiplexed so that the function is not completely lost even if a failure is happened in one of the detectors.

The safest approach for safety channel is to scram a reactor whenever an anomaly is detected, even if it is attributable to a false detection. This kind of approach is called "fail safe," which is adopted for low-power reactors that are easy to start up and shut down. On the other hand, in a nuclear power plant or a high-power research reactor, if the reactor is shut down due to a false detection, it takes a long time to restart, resulting in a significant loss of availability of the reactor, significant economic and social impact. Therefore, it is important to keep the reliability of the safety channel high. A typical example of such a system is the "2 out of 3" system that three detector systems are prepared for the same measurement target. Of three systems, when only one system detects an anomaly, the event is regarded as a false detection. Conversely, when two or more systems detect simultaneously an anomaly, the event is regarded as an anomaly.

UTR-KINKI has two independent and identical safety channels, and each channel uses the UIC in the current mode as a neutron detector. The current signal from the UIC is converted to a voltage signal and displayed as power in percentage on the safety channels #1 and #2 in the control console. The UIC is chosen for the simplest structure and principle among neutron detectors and is expected to have long-term stability and high reliability. Both channels will generate a scram signal to shut down the reactor, when the reactor power exceeds 150% (1.5 W) of the licensed power.

Bibliography

Comprehensive explanations (in Japanese) of nuclear instrumentation of research reactors

1 Kawaguchi T, Ara K (1978) Genshiro no keisoku (Measurement of nuclear reactor). Saiwai Shobo, Tokyo, Japan (in Japanese)
2 Sumita K, Katoh K, Furukawa T, Wakayama N (1976) Genshiryoku keisoku (Nuclear instrumentation). Nikkan Kogyo Shimbun, Ltd., Tokyo, Japan (in Japanese)
3 ATOMICA Nuclear reactor measurements (1) process instrumentation (03-06-05-01). (in Japanese) https://atomica.jaea.go.jp/data/detail/dat_detail_03-06-05-01.html. Accessed 1 July 2022
4 ATOMICA Nuclear reactor measurements (2) nuclear instrumentation (03-06-05-02). (in Japanese) https://atomica.jaea.go.jp/data/detail/dat_detail_03-06-05-02.html. Accessed 1 July 2022

Detection principles of neutron detectors

5 Knoll GF (2010) Radiation detection and measurement, 4th edn. Wiley, Hoboken, NJ

Chapter 6
General Aspects of Laws and Regulations Relating to Nuclear Energy in Japan

Abstract The International Atomic Energy Agency has established Fundamental Safety Principles, which state that nuclear safety and security measures should be aimed at protecting human life and health and the environment. Based on the universal principles, it is the responsibility of each country to establish safety regulations for the peaceful use of nuclear energy, and, in Japan. In Japan, "Act on the Regulation of Nuclear Source Material, Nuclear Fuel Material and Reactors (hereinafter, referred to as "Act on the Regulation of Nuclear Reactors")" has been established. Also, compliance with this act and related laws and regulations is required for the establishment and operation of nuclear facilities. This chapter provides an overview of Japanese laws and regulations related to the peaceful use of nuclear energy.

Keywords Regulation · Act · Low · Nuclear safety · Nuclear security · Safeguards

6.1 Legal System and Position of Laws and Regulations

The use of nuclear energy and radiation is regulated by various laws and regulations based on Atomic Energy Basic Act. Figure 6.1 shows the hierarchical classification system of laws in Japan. Acts are enacted by the Diet, which is a legislative body. Since the matters prescribed by Acts are, however, limited to fundamentals, there are "orders" as norms established by the administrative office to supplement the acts. The orders include the "Cabinet Order" enacted by the Cabinet and the "Ordinance for Enforcement" established by the relevant regulatory authorities. In addition, there are "guides" and "notices" that are not laws and regulations but are issued by regulatory authorities to achieve the purposes of laws and regulations. These are generally actions of informing specific persons or unspecified large numbers of persons of specific matters and include those indicating the operation of laws and administrative enforcement policies. When it is necessary to inform the public of the contents of a notice, it is published in the official gazette as a "public notice".

There are 12 acts related to the use of atomic energy. As for radiation safety for workers is regulated by the Industrial Safety and Health Act under the jurisdiction of the Ministry of Health, Labour and Welfare (MHLW), medical use is regulated by the Medical Care Act and Act on Securing Quality, Efficacy and Safety of Products

© The Author(s) 2023
G. Wakabayashi et al., *Introduction to Nuclear Reactor Experiments*,
https://doi.org/10.1007/978-981-19-6589-0_6

Act: A national norm enacted by a resolution of the Diet in accordance with the Constitution

Cabinet order: Enacted by the Cabinet (administrative organization). Based on the provisions of laws and government ordinances, it is delegated by the laws to provide for details that are not provided for in the laws, mainly concerning specific procedures.

Ministerial ordinance: Establishes specific matters for implementing laws based on the provisions of laws and government ordinances enacted by the relevant administrative organs.

Fig. 6.1 Hierarchical classification system of laws

Including Pharmaceuticals and Medical Devices, and transportation of radioactive materials is regulated by various laws and regulations under the jurisdiction of the Ministry of Land, Infrastructure, Transport and Tourism (MLIT). The MLIT has jurisdiction over the transport of radioactive materials.

6.2 Atomic Energy Basic Act

The Atomic Energy Basic Act is the Act that sets forth the basic policy of Japan's nuclear energy policy. The Atomic Energy Basic Act was enacted on December 19, 1955, for the purpose of securing energy resources for the future, promoting academic progress and industrial development and thereby contributing to the welfare of human society and improvement of the standard of living of the people by promoting research, development and utilization of nuclear energy (hereinafter referred to as "utilization of nuclear energy") (Article 1 of the Act). Its basic policy is set forth in Article 2 as follows:

- Article 2 The research, development and utilization of nuclear energy is limited to peaceful purposes, is to aim at ensuring safety, and is performed independently under democratic administration, and the results obtained is made public so as to actively contribute to international cooperation.

- The security set forth in the preceding paragraph shall be carried out in accordance with established international standards and with the aim of contributing to the protection of the lives, health and property of the citizens, the preservation of the environment and the security of Japan.

This basic policy is based on the following principles:

(1) Information on research, development and utilization of nuclear energy shall be fully disclosed and made known to the public.
(2) Nuclear research should be democratically administered and require the full cooperation of all competent researchers.
(3) The research and use of nuclear energy should be conducted under autonomous management.

The three principles of nuclear power in Japan are "democracy," "independence" and "openness." The three principles were also proposed at the 17th general meeting of the Science Council of Japan (April 1954).

The Nuclear Regulation Authority (NRA) and its secretariat, the Nuclear Regulation Agency, were established in 2012 in response to the accident at TEPCO's Fukushima Daiichi Nuclear Power Station in 2011, and are responsible for ensuring safety in the use of nuclear energy by exercising their authority in a neutral and fair manner independent of the government. The NRA was established as an external organization of the Ministry of the Environment (MOE) by separating the safety and regulatory divisions from the organizations in charge of research, development and utilization of nuclear energy, such as the Ministry of Education, Culture, Sports, Science and Technology (MEXT) and the Ministry of Economy, Trade and Industry (METI).

The Atomic Energy Basic Act states that the management of nuclear fuel materials and nuclear reactors shall be subject to government regulations as prescribed separately by law (Article 14), and the Nuclear Reactor Regulation Law has been enacted. In addition, regulations on the manufacture, sale, use and measurement of radioactive materials and radiation generating devices, and other safety and health measures shall be prescribed by law (Article 20 of the Atomic Energy Basic Act), and the Radioisotope Regulation Act has been separately established.

In addition, the following five items are defined as applicable to the Atomic Energy Basic Act and related laws and regulations (Article 3 of the Atomic Energy Basic Act).

(1) The term "nuclear energy" means all types of energy emitted from the nucleus of an atom in the process of nuclear transmutation.
(2) The term "nuclear fuel material" means materials that emit high energy in the process of nuclear fission, such as uranium and thorium, which are specified by a Cabinet Order.
(3) The term "nuclear source materials" means materials that are used as the raw materials of nuclear fuel materials, such as uranium ore and thorium ore, which are specified by a Cabinet Order.

Table 6.1 Nuclear fuel materials requiring permission for use regardless of quantity

	Type
1	Uranium (U) and its compounds in which the ratio of U-235 to U-238 exceeds the natural mixing ratio
2	Plutonium (Pu) and its compounds
3	U-233 and its compounds

(4) The term "reactor" means a device that uses nuclear fuel materials as fuel; provided, however, that those specified by a Cabinet Order.

(5) The term "radiation" means electromagnetic waves or particle beams capable of ionizing air directly or indirectly that are specified by a Cabinet Order.

The scope of application of relevant laws and regulations is determined by the five sections mentioned above. The cabinet order to which specified by "cabinet order refers" is the cabinet order concerning the definition of nuclear fuel material, nuclear material, nuclear reactor and radiation (hereinafter referred to as the definition by the cabinet order). According to the definitions by the cabinet order, the items excluded from "reactors" shown in Section (4) are items other than devices that can control fission chain reactions and can sustain, or are likely to sustain, the equilibrium state of reactions without using a neutron source. In other words, there is no exemption level based on the power of the reactor, and UTR-KINKI, which is a reactor with a thermal power of 1 W, is regulated as a "reactor" under the Atomic Energy Basic Act and related laws and regulations. As for Item (2), there are two types of nuclear fuel materials: one that requires a license regardless of the quantity, and the other that does not require a license for use within a specified quantity. Each of them is shown in Tables 6.1 and 6.2. Nuclear material is defined as "any material containing uranium or thorium or their compounds other than nuclear fuel material." The definition of "radiation" in Section (5) is also shown in Table 6.3.

6.3 Act on the Regulation of Nuclear Reactors, etc.

Act on the Regulation of Nuclear Reactors, etc., abbreviated as "Act on the Regulation of Nuclear Source Material, Nuclear Fuel Material and Reactors," was enacted in 1957 as a law to regulate the handling of nuclear reactors and nuclear materials in general, based on the Atomic Energy Basic Act. The purpose of this law is as follows:

- Ensuring that the use of nuclear source materials, nuclear fuel materials and nuclear reactors is limited to peaceful purposes.
- In the event of a severe accident at a nuclear facility, preventing the release of radioactive materials at an abnormal level outside the plant or place of business where said nuclear facility is established and other disasters caused by nuclear source materials, nuclear fuel materials and nuclear reactors.

Table 6.2 Types and quantities of nuclear fuel materials for which permission for use is not required

	Type	Quantity for which permission for use is not required
1	U and its compounds in which the ratio of U-235 to U-238 is a natural mixing ratio	U content of 300 g or less
2	U and its compounds where the ratio of U-235 to U-238 does not reach the natural mixing ratio	U content of 300 g or less
3	Substances containing one or more of Types 1 and 2 that can be used as fuel in a nuclear reactor	U content of 300 g or less
4	Thorium (Th) and its compounds	Amount of Th less than 900 g
5	Substances containing one or more of the following 4 that can be used as fuel in a nuclear reactor	Amount of Th less than 900 g

Table 6.3 Radiation in Article 3, Section (5) of the basic act on nuclear energy

1	Alpha rays, deuteron rays, proton rays and other heavy charged particle rays and beta rays
2	Neutron beam
3	Gamma rays and characteristic X-rays (limited to characteristic X-rays generated by orbital electron capture)
4	Electron beams and X-rays with an energy of 1 meV or more

- To protect nuclear fuel materials and to ensure public safety, necessary regulations shall be imposed on the smelting, processing, storage, reprocessing and disposal businesses and the establishment and operation of nuclear reactors, taking into account the occurrence of large-scale natural disasters, terrorism and other criminal acts.
- To implement the convention on the research, development and use of nuclear energy and other international commitments by imposing necessary restrictions on the use, etc. of internationally controlled materials, thereby contributing to the protection of the life, health and property of the citizens, the preservation of the environment and the security of Japan.

Act on the Regulation of Nuclear Reactors, etc. was tightened after the accident at the Fukushima Daiichi Nuclear Power Plant in March 2011, and the so-called "new regulatory standards" were introduced. One of the problems with the safety regulations prior to the accident at the Fukushima Daiichi Nuclear Power Plant was that severe accident countermeasures were not regulated and there was insufficient preparation. In addition, there was no legal system to apply the new standards retroactively to existing nuclear power plants, resulting that the highest level of safety was not always ensured. In response to this shortcoming, "thoroughness of deep protection" was introduced as a basic concept of the new regulatory standards. This is intended to prepare multiple (multilayered) countermeasures that are effective in achieving the objectives, and when considering countermeasures for each layer, not to expect countermeasures in other layers. The intension required the further increase in the assumptions related to natural phenomena and the strength of protective measures against them, in order to prevent the simultaneous loss of safety functions due to common factors. In addition, it was also required to strengthen the countermeasures against the loss of safety functions simultaneously due to common factors other than natural phenomena, such as fire, internal water overflow in the facility and power failure. Specific measures to meet the new regulatory standards will be selected by operators according to the characteristics of each facility. The newly strengthened safety measures were applied not only to nuclear power plants but also to test and research reactors and nuclear fuel cycle facilities. Furthermore, one of the features of the new regulatory standards is that, in the event that new findings are obtained and the licensing standards are changed, or in the event that the licensing standards are not changed but the reactor facility no longer conforms to the licensing standards, the NRA may request the establisher of the reactor, etc. concerned to suspend the use of the facility, remodel, repair or relocate the facility, designate the method of operation of the reactor or take other necessary measures for safety. This is called a "backfit order." As a result, all nuclear reactors in Japan have been shut down after the enforcement of this law and have been restarted one after another, starting with those that have been approved to comply with the new regulatory standards.

To evaluate the adequacy of the design, a safety assessment is performed assuming the occurrence of several "design basis events." A safety evaluation is performed by assuming the occurrence of several design basis events. Such an event is generally called a severe accident [1]. In the previous regulations, standards to prevent such severe accidents (design standards) were established, and it was confirmed that the failure of a single component would not lead to core damage. Here, the new regulatory standards require measures to stop the development of severe accidents and to deal with intentional aircraft collisions as a form of terrorism, on the basis of the additional assumption of the occurrence of terrorism to the purpose of the law. Figure 6.2 shows a comparison between the existing regulatory standards and the new regulatory standards for commercial power reactors. As shown in Fig. 6.2, many requirements have been newly established and strengthened. Under the new regulatory standards, investigations of active faults and underground structures are required again, and the reference earthquake ground motions and seismic strengthening are being revised as necessary. With regard to tsunami, which is a factor in the accident at the Fukushima Daiichi Nuclear Power Plant, the location and height of their occurrence have been assessed, and measures have been taken to ensure the functioning of safety–critical equipment, the installation of breakwaters and seawalls, and the making of doors watertight. In addition, since countermeasures against volcanoes, tornadoes, forest fires, etc. are newly required, these effects on the safety of nuclear power plants need to be evaluated and countermeasures be taken as necessary. In test and research reactors, these measures need to be taken under the new regulatory standards basically in the same way as for commercial nuclear power plants.

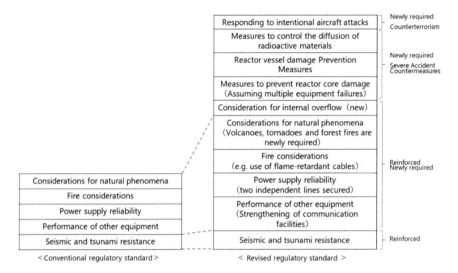

Fig. 6.2 Comparison of conventional regulatory standards and new regulatory standards (Ref. [2]) (Translated under the creative commons attribution license 4.0. (https://creativecommons.org/lic enses/by/4.0/) from [2] Copyright © nuclear regulation authority)

6.4   Nuclear Security and Physical Protection

Nuclear security refers to the measures, ensuring that possible threats from the misuse of nuclear materials or radiation sources do not materialize. These principles are common throughout the world, and a framework for ensuring international nuclear safety has been developed under the International Atomic Energy Agency (IAEA). The IAEA has established its Fundamental Safety Principles (SF-1) as the measures for nuclear safety and security, which aim to protect human life, health and the environment. Nuclear security refers to protective measures, such as countermeasures against nuclear terrorism, not only for nuclear material but also for all radioactive materials including radioisotopes. In Japan, security measures for nuclear materials are stipulated in the Act on the Regulation of Nuclear Reactors, etc., and for radioisotopes, security measures are stipulated in the Act on the Regulation of Radioisotopes, etc., mainly to prevent theft and theft. Specifically, the following four threats, which are assumed to be the misuse of nuclear materials and radioisotopes by terrorists, etc., have been identified: (1) theft of nuclear weapons; (2) manufacture of nuclear explosive devices using the stolen nuclear materials; (3) manufacture of devices to disperse radioactive materials (so-called dirty bombs); (4) sabotage of nuclear facilities and transportation of radioactive materials. Physical protection refers to measures taken to protect nuclear materials and facilities from sabotage and vandalism, and from theft and sabotage of nuclear materials in use, storage, and transport.

Nuclear material protection mainly applies to Pu and enriched U, and the necessary protection requirements are classified into three levels according to the type and amount of nuclear material. The nuclear material subject to these regulations is called "specified nuclear fuel material".

6.5 Safeguards

In 1970, the Treaty on the Non-Proliferation of Nuclear Weapons (NPT) came into force with the aim of reducing the possibility of nuclear war by preventing the number of nuclear weapon states (the five countries that possessed nuclear weapons as of January 1, 1967: the USA, the former Soviet Union, the UK, France and China) from increasing further. Japan signed the treaty on February 3, 1970, and ratified it on June 8, 1976. The treaty stipulates that "Japan shall conclude an agreement with the IAEA in accordance with the IAEA's safeguards system, and the parties shall accept the safeguards stipulated in the agreement," and Japan has accepted the safeguards in accordance with the agreement.

The acceptance of the agreement has a great influence on the test and research reactors. Since the purpose of the test and research reactors is to utilize neutrons, "highly enriched uranium (HEU)" has been used to obtain high power density and to increase neutron flux in a small core. However, due to the nuclear non-proliferation policy of the USA, the supplier countries are required to make efforts to reduce fuel enrichment as a condition of uranium supply. Under the initiative of the USA, the low enrichment of the fuel of the test and research reactors has obtained an international consensus, and the low enrichment of the fuel of test and research reactors has also been promoted in Japan [3].

6.6 Act on the Regulation Radioisotopes, etc.

Act on the Regulation of Radioisotopes, etc. was newly enacted on September 1, 2019. It was formerly known as the "Act on Prevention of Radiation Hazards due to Radioisotopes, etc." The purpose of the new Act is to prevent radiation hazards caused by radioisotopes and to ensure public safety by regulating the use, sale, lease, disposal and other handling of radioisotopes, the use of radiation generating equipment, and the disposal and other handling of radioisotopes or objects contaminated by radiation generated by radiation generating equipment (hereinafter referred to as "radioactive contaminated objects") in accordance with the spirit of the Atomic Energy Basic Act, to prevent radiation hazards and to ensure public safety. Subsequently, the regulations were reviewed mainly to conform to the 2011 Nuclear Security Recommendations on Radioactive Materials and Related Facilities, and the name of the law was changed to define the "specific radioisotopes" that are particularly dangerous and to add the protection of radioisotopes to the purpose of the law. The use of radionuclides and radiation generating devices other than nuclear fuels and materials regulated by the Act on the Regulation of Nuclear Reactors, etc. requires permission (notification) to be obtained as stipulated by the relevant laws. Radioisotopes, instruments equipped with radioisotopes and radiation-producing devices are exempted from regulation if the quantity is less than the minimum quantity specified in the public notice. In addition, various limits for radiation doses, etc. are specified in this law. Major regulatory values are shown in Table 6.4 and Fig. 6.3. The boundary of the plant in Fig. 6.3 corresponds to the "environmental monitoring area" that is defined as the area around the controlled area in the Act on Regulation of Nuclear Reactor. The controlled area is defined as an area where dose limits are not likely to be exceeded at any place outside the area, and the dose limits are the same as the dose limits at the boundaries of business establishments in the Act on the Regulation Radioisotope, etc.

Table 6.4 Occupational dose limits

Item	Occupational dose (Radiation worker)
Effective dose limits	(1) 100 mSv/5 years (2) 50 mSv/year (3) Women 5 mSv/3 months (4) Pregnant women 1 mSv (internal exposure) for a period of time starting from the time when her employer and others are informed of her pregnancy by her reporting or any other means up to the time of the delivery of the baby
Equivalent dose limits	The lens of the eye (1) 100 mSv/5 years (2) 50 mSv/year Skin 500 mSv Pregnant women 2 mSv for abdomen surface (up to the time of the delivery of the baby)

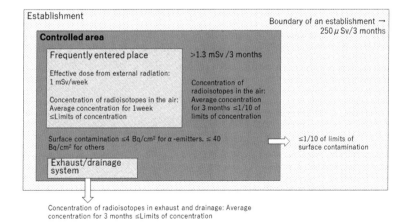

Fig. 6.3 Maximum permissible radiation level to be maintained in and around controlled area

6.7   Ordinance on Prevention of Ionizing Radiation Hazards

Nuclear energy and radiation are used in various situations such as medical facilities in addition to nuclear reactor facilities and nuclear fuel facilities. For such workers who may be exposed to radiation in the course of their occupation, employers need to make efforts to minimize their exposure as much as possible. This basic principle is set forth in the Regulations for Prevention of Ionizing Radiation Hazards. These rules are designed to ensure the safety and health of workers in the workplace and to promote the formation of a comfortable working environment by promoting comprehensive and systematic measures for the prevention of occupational accidents, including the establishment of hazard prevention standards, clarification of responsibility systems, and measures to promote voluntary activities. The rules are also based on the provisions of the Industrial Safety and Health Law and the Order for Enforcement of the Industrial Safety and Health Law.

Radiation work to which this regulation applies is defined in the Ordinance for Enforcement of the Industrial Safety and Health Act (Appended Table 2 Radiation Work) as follows:

(1)   Work to use X-ray equipment or inspection of the said equipment that involves the generation of X-rays

(2)   Work to use a cyclotron, betatron or other device for generating charged particles or to use ionizing radiation (meaning alpha rays, deuteron rays, proton rays, beta rays, electron rays, neutron rays, gamma rays and X-rays; the same shall apply in Section (5))

(3)   Work to vent X-ray tubes or kenotrons, or work to inspect these involving the generation of X-rays

(4)   Work to handle equipment with radioactive materials specified by an Ordinance of the Ministry of Health, Labour and Welfare

(5)   Work to handle radioactive materials prescribed in the preceding item or objects contaminated by ionizing radiation generated from said radioactive materials or equipment prescribed in Section (2)

(6)   Work to operate nuclear reactors

(7)   Work to mine nuclear source materials (which means nuclear source materials prescribed in Article 3, Section (3) of Atomic Energy Basic Act (Act No. 186 of 1955)) in a mine

In addition to these types of work, there is work to decontaminate soil contaminated by radioactive materials originating from the accident at the Fukushima Daiichi Nuclear Power Plant. The "Ordinance on Prevention of Ionizing Radiation Hazards pertaining to Work to Decontaminate Soil, etc. Contaminated by Radioactive Materials Caused by the Great East Japan Earthquake (Ministry of Health, Labour and Welfare Ordinance No. 152, 2011)" which stipulates the basic principles for the prevention of radiation hazards pertaining to this work is separately established.

In addition to the exposure dose limits for workers, the Ionization Ordinance stipulates the obligations of business operators, such as working environment measurement, medical examinations of workers and education. With regard to medical examinations, workers who are constantly engaged in radiation work and who enter the controlled area are required to undergo medical examinations by a physician with regard to the specified items on a regular basis at the time of employment or reassignment to the relevant work and once every six months thereafter. Similar provisions to the Ionization Measurement are also found in the Act on the Regulation Radioisotope, etc. and other laws. Since the Ordinance on Prevention of Ionizing Radiation Hazards is for workers, student radiation workers are subject to the provisions of the Act on the Regulation Radioisotope, etc.

6.8 Nuclear Disaster Prevention

When an accident or an abnormal event occurs in a nuclear facility, radioactive materials can be released, which may directly or indirectly cause damage to the residents and the environment in the surrounding area of the facility. It is necessary to take countermeasures against such a nuclear disaster with regard to radioactive materials and radiation. For this purpose, the national government and local governments set up monitoring posts outside the nuclear facility site. Governments determine the actions to be taken to avoid exposure of residents (protective measures) based on the measured values of radiation levels (ambient dose rate and radioactivity concentration in the air), etc. The "Act on Special Measures Concerning Nuclear Emergency Preparedness" has been established to deal with such a nuclear disaster. In view of the unique nature of a nuclear disaster, the Act stipulates special measures to be taken by nuclear operators in relation to the prevention of a nuclear disaster, the issuance of a declaration of a nuclear emergency, the establishment of nuclear emergency response headquarters, the implementation of emergency response measures and other matters related to a nuclear disaster. The purpose of the Act is to protect the lives, bodies and properties of the citizens from a nuclear disaster by strengthening measures against a nuclear disaster (Article 1 of the Act).

Based on the lessons learned and experiences gained from the Fukushima Daiichi Nuclear Power Plant accident, the disaster prevention system for nuclear disaster has been reviewed by revising the Act on Special Measures Concerning Nuclear Emergency Preparedness and related laws and regulations. In the event of a nuclear disaster, the Nuclear Emergency Response Headquarters, headed by the Prime Minister, is established in the Prime Minister's Office. Under the Act on Special Measures Concerning Nuclear Emergency Preparedness, the Nuclear Regulatory Commission (NRC) is to establish "Guidelines for Nuclear Emergency Preparedness." The accident at the Fukushima Daiichi Nuclear Power Station has revealed that there are many problems with the existing disaster prevention measures for nuclear facilities. Based on the issues revealed, the "Guidelines for Nuclear Emergency Preparedness" were formulated based on the following basic concepts:

- Formulate a disaster prevention plan from the perspective of residents.
- To build a system that can provide information on a continuous basis.
- The latest international findings should also be incorporated, and the criteria for decisions used in planning should be constantly reviewed to ensure that they are optimal.

The Cabinet Office is in charge of nuclear emergency preparedness. Specifically, the Cabinet Office supports the preparation of regional disaster prevention plans and evacuation plans, enhances the nuclear emergency preparedness system, provides financial support to related prefectures and conducts trainings.

The current approach to nuclear emergency preparedness has been revised in the wake of the accident at the Fukushima Daiichi Nuclear Power Plant. For example, off-site evacuation is divided into Precautionary Action Zone (PAZ) and Urgent Protective action planning Zone (UPZ) according to the distance from the nuclear facilities, as shown in Fig. 6.4, based on the IAEA standards. As shown in Fig. 6.4, two emergency zones, PAZ and UPZ, are defined in advance and evacuation procedures are considered involving evacuation procedure. In the case of a commercial power reactor, the PAZ is defined as a zone within a distance of approximately 5 km, where precautionary evacuation measures are taken before radioactive materials are released. The UPZ is defined as a zone of approximately 5 km to 30 km, where people are first evacuated indoors. If a certain amount of air radiation is measured after the release of radioactive materials, the area will be identified and temporary relocation or evacuation will be carried out sequentially.

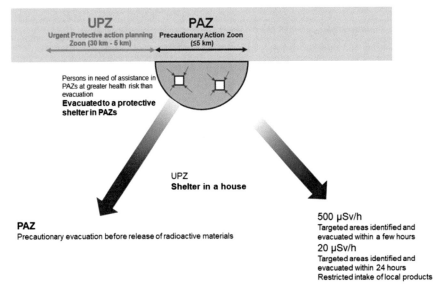

Fig. 6.4 Concept of off-site evacuation based on nuclear emergency response guidelines (Ref. [4]) (Translated under the creative commons attribution license 4.0 (https://creativecommons.org/lic enses/by/4.0/) from [4] Copyright agency for natural resources and energy)

Each nuclear operator shall establish a "nuclear operator emergency action plan" based on the Act on the Prevention of Nuclear Emergency Preparedness and clarify matters concerning necessary tasks to be carried out by nuclear emergency preparedness managers, in order to prevent the occurrence and progression of a nuclear disaster and to achieve restoration from a nuclear disaster, including measures to prevent nuclear emergency, emergency response measures and measures after a nuclear disaster. Each operator shall take the following measures based on the Nuclear Emergency Preparedness Plan, each operator has been promoting safety measures to prevent the occurrence of a nuclear disaster, and has been working on measures for nuclear emergency preparedness, such as continuous implementation of trainings to improve response capabilities in case of a severe accident and strengthening of cooperation with relevant local governments, so that emergency facilities can be effectively utilized in the event of a nuclear disaster.

References

1. Safety in nuclear safety: severe accidents. http://anzenmon.jp/vizwik/app/view_page_printable. html;jsessionid=92CFF12736DBC01AFA7B612D5B68A9CD?id=6948. Accessed 1 July 2022
2. Nuclear regulation authority of Japan, new regulatory standards for commercial power reactors. (in Japanese) https://www.nsr.go.jp/data/000070101.pdf. Accessed 1 July 2022
3. ATOMICA research reactor fuel reduction project (REPTR) (03-04-01-04). (in Japanese) https://atomica.jaea.go.jp/data/detail/dat_detail_03-04-01-04.html. Accessed 1 July 2022
4. Agency for natural resources and energy, emergency preparedness for nuclear emergency. (in Japanese) https://www.enecho.meti.go.jp/about/special/johoteikyo/genshiryokubousai.html. Accessed 1 July 2022

Further Readings

5. Federation of electric power companies of Japan, INFOBASE j-9 electric utility database (INFOBASE). https://www.fepc.or.jp/library/data/infobase/pdf/06_j.pdf. Accessed 1 July 2022

Printed in the United States
by Baker & Taylor Publisher Services